Congratulations!

With your purchase of the *Professional Writing Online Users' Handbook*, Longman Publishers is pleased to provide you with a twelve-month subscription to the Professional Writing Online Website, a unique set of educational materials designed to be used in your course.

To explore this valuable learning resource:

1. Go to www.ablongman.com/pwo
2. Click on the Professional Writing Online icon
3. Follow the instructions to register your subscription using the following access code:

        WSPO-CHESS-THOLE-PICON-FINIS-LAIRD

4. During the registration process, you will choose a personal user ID and password for use in logging into the site. Once your subscription is confirmed, you can begin using Professional Writing Online immediately.

Your pre-assigned code can be used only once to establish your subscription, which is not transferable and is limited to twelve months from the date of activation. If you purchased a used *Professional Writing Online Users' Handbook*, this access code may have already been used. However, you can find information on how to purchase a subscription at the Professional Writing Online registration page.

http://www.ablongman.com/pwo

Welcome to Professional Writing Online, a new online resource for students and instructors of workplace writing courses. The pedagogical mission of Professional Writing Online is to provide course materials in a lively, dynamic, and engaging fashion. The website offers a variety of learning tools and references designed to make study and writing easier, more engaging, and more rewarding.

Mail: Pearson Higher Education
Central Media Group - Product Support
1900 East Lake Avenue
Glenview, IL 60025

Phone: (800) 677-6337 (Mon. to Fri., 8:00 a.m. to 5:00 p.m. CST)
Fax: (847) 486-3698
E-mail: webtech.support@pearsoned.com
WWW: http://www.ablongman.com/techsupport/

Professional Writing *Online*

Users' Handbook

James E. Porter
Case Western Reserve University

Patricia Sullivan
Purdue University

Johndan Johnson-Eilola
Clarkson University

Allyn and Bacon
Boston London Toronto Sydney Tokyo Singapore

Editor-in-Chief: Joseph Opiela
Cover Designer: Nancy Danahy
Technical Desktop Manager: Heather A. Peres
Manufacturing Buyer: Lucy Hebard
Printer and Binder: R.R. Donnelley & Sons
Cover Printer: Phoenix Color Corp.

Copyright © 2001 by Allyn & Bacon.

All rights reserved. No part of this publication may be reproduced, stored in a retrieval system, or transmitted, in any form or by any means, electronic, mechanical, photocopying, recording, or otherwise, without the prior written permission of the publisher. Printed in the United States.

Please visit our website at http://www.ablongman.com/pwo

ISBN 0-205-27918-X

2345678910-DOC-030201

table of

Contents

Welcome to Professional Writing Online 1

    How to Use Professional Writing Online 1
    Community and Interactivity 2
    Using the Handbook 2
    Acknowledgements 3

Chapter 1: The Project-Oriented Approach 5

    Scenario: Teaching from a Project-Oriented Approach 6
    Scenario: Teaching from a Document or Principles Oriented Approach 7
    Scenario: Using PWOnline in the Distance Education Course 8
    Scenario: Using PWOnline in Traditional Classrooms 9
    Scenario: Tailoring Projects for Local Situations 10
    Scenario: Developing Distance Collaboration 10
    Scenario: Using PWOnline to Support International Understanding 11

Chapter 2: Site Map 13

    Table 2-1: Professional Writing Online Sitemap 15

Chapter 3: Projects 17

    Table 3-1: Projects in Professional Writing Online 17
    Software Learning Initiative 19
    Designing Online Documentation 21
    Evaluating the Design of Online Documentation 24
    Analysis of Writing Practices 26
    Analyzing Professional Contexts 28
    Customer Complaints at Allied Mutual 32
    Airbag Safety and Customer Service at Vanguard Motor Company 36
    Copyshop Copying Policy Case 38
    United Drill Case 40
    The Corporate Web Project 42

The Big 1 Case	45
The Kmart Case, or the Renegade Webmaster	48
Employment Project	50
The Lester Crane Case	51
Technology Access Memo/Report for Distance Education Course	56
NCAA Baseball Bat Standards: Complaints and Policy	59
Software Bug Case	62
The Job Banks Project	65
National Electric: Creating Visuals for Employee Statistics	68
E-Commerce Project	70

Chapter 4: Documents — 73

Sample Letter - Request for Deposit Refund	74
Sample Letter - Revised Request for Deposit Refund	76
Sample Memo - Announcement of Policy Change	78
Sample Memo - Revised Announcement of Policy Change	79
Sample Memo - Project Planning Memo	81
Sample Memo - Revised Project Planning Memo	83
Sample Report - Corporate Web Project Proposal	86
Sample Report - Corporate Web Recommendation Report	89
Sample Report - Documentation Project Plan	93
Sample Application Letter - Research Position	99
Sample Resume - Construction Management Internship	101
Sample Resume - Construction Internship (detail)	103
Sample Reference Page	105

Chapter 5: Principles — 107

Contents of "Principles"	108
The Principled Perspective	112

Chapter 6: Resources — 113

Writer's Web Links	113
Writer's Exercises and Worksheets	114
Instructor's Resources	116
Interactive Features and Functions	116
Feedback	119

About the Authors — 121

Index — 123

welcome to

Professional Writing Online

How to Use Professional Writing Online
Some Advice from the Authors

Professional Writing Online (or PWOnline for short) is an instructional web site — a "web textbook," if you will. Any teacher, student, or workplace writer with access to the World Wide Web will be able to call up this web site and use it as a resource for teaching and learning professional writing.

The web site is located at:

<http://www.ablongman.com/pwo>

We have designed PWOnline primarily for use by students and instructors in university-level business and technical writing service courses and in major courses in professional/technical writing. It could be used by teachers in a traditional classroom. It could also be used for corporate training or personal learning. However, its primary use is as a learning environment for distance education courses and for professional writing courses taught in computer classrooms.

PWOnline is not a book — and it's probably best not to try to use it like one. We have been using PWOnline as a teaching resource for approximately two years now — and for the past year (1999-2000) it has been the principle instructional text used in computer-based technical and business writing courses at both Purdue University and Case Western

Reserve University. We have found PWOnline to be an effective tool in those courses; however, like most tools, its effectiveness is determined largely by its situated use and application. (A hammer is a great tool for driving nails, not so good for washing windows.) Below we provide some general advice about how PWOnline might be used for maximum effectiveness.

Community and Interactivity

One helpful quality of PWOnline, we hope, will be its interactive quality. PWOnline provides discussion lists and a chat space for students, teachers, and workplace professionals to interact and discuss matters of common interest related to professional communication. A teacher discussion list will allow instructors using the site to discuss its use, to compare notes, and to recommend changes and updates. Students will be able to chat with other students and with teachers about the projects they are working on.

We have plans to expand the interactivity and community nature of PWOnline. For instance, we are hoping that instructors will contribute sections to the site — or at least allow links from the site to their own instructional material.

Far more than a textbook, PWOnline is an instructional environment, a work space, an online community of learners. We hope that PWOnline can become a meeting place for professional writers.

Using the Handbook

The *Professional Writing Online Users' Handbook* is the print guide to the PWOnline web site. It is not an independent textbook. It serves as a users' manual for the web site.

We have included in this *Users' Handbook* material that we think will help a variety of folks benefit from *Professional Writing Online*. The handbook provides an overview of the site and advice about how to use the site, but it also includes some content from the site that writers and instructors might find most useful to have in print form. For example, we have provided some sample documents of the various genres of business and technical communication, and we have printed out the introductory description for each of the major projects in the site. (Our experiences as teachers using these materials in business and technical communication courses suggest that students and teachers most want print versions of these sorts of materials. Conversely, they only rarely and selectively print out discussions of principles.)

We hope this handbook and the Professional Writing Online web site provides the advice, the samples, and the assistance you need to become a better professional writer, whether you are a teacher, a student, a professional writer or trainer, or an independent learner. That is the main goal of *Professional Writing Online*: to help professional writers (and professional writing instructors) do their work more effectively, more productively, more successfully.

The Authors — Jim, Pat, Johndan

Acknowledgements

The teachers of English 420 and English 421 at Purdue University were invaluable to our work. In particular we wish to note the program assistants and mentors of recent years:

    Gary Beeson
    Kristine Blair
    Teena Carnegie
    Bill Hart-Davidson
    Theresa Fishman
    Tim Krause
    Elizabeth Lopez
    Tim Peeples
    Colleen Reilly
    Michel Simmons
    Ann Maire Simpkins

chapter 1

The Project-Oriented Approach

As a web site, PWOnline can be used in many different ways — and that is its chief advantage over conventional print textbooks. It offers four primary points of entry, four major sections:

- Projects
- Documents
- Principles
- Resources

A teacher who wants to employ a situation-based approach in the classroom would start a course in the Projects section. A busy employee at work who wants some quick advice on how to write a memo might begin in the Documents section, locate a sample memo template, and then link to the subsection providing a discussion of the rhetoric of the memo as well as sample memos (with commentary). A teacher looking for sample syllabi for a distance education course in technical writing would go to the Resources section and find syllabus links there. Of course, the sections are interlinked — so that students and teachers can move back and forth across sections to find what they need. As a web site, PWOnline provides far more information than the conventional print textbook — and of course it has links to the many resources available on the World Wide Web.

As teachers of technical and business writing/communication, we prefer a project-oriented approach. We tend to begin each assignment and each instructional module in the Projects section of PWOnline — and then move from there to Documents, Principles, Resources (often in this order). Why this approach? A project-oriented approach is based on a rhetorical view of communication that sees purpose and audience as the starting point, the driving impetus behind communication acts. We feel that real-life communication situations are driven by situations — that is, scenarios that create a need for writing or communication. Yes, sometimes genre seems to drive the need for writing: a technical writer can be told to "submit a progress report" or an information developer can be asked to "create a web site." It is certainly vital that professional writing students learn the various genres of professional writing. However, genres exist to serve organizational and human needs. Genres are themselves conventional forms that have arisen because they serve the needs of "repeated situations." The progress report is necessary so that management can keep track of a project history and evaluate its progress and success. The web site is necessary because a company needs to compete in the virtual marketplace through developing e-commerce and electronic recruiting initiatives. In situations, the genres serve organizational needs.

There is a pedagogical as well as theoretical rationale to favoring the project-oriented approach. In our experience as teachers, we find that students seldom see Principles as interesting or useful by themselves or in the abstract. When they are immersed in a concrete task or situation, students can more clearly see that the principles have an application and a usefulness; they can also see that principles have limitations, and they start to engage the principles by disagreeing with them or complicating them. At that point, understanding happens — as students begin to develop their own version of the principles based on their own communication experiences.

Here is an example of how the project-oriented approach works:

Scenario: Teaching from a Project-Oriented Approach

When Jim teaches business communication, he typically starts the class with a short case, such as the Allied Insurance case. This case is designed to introduce students to the rhetorical basics of professional communication: Who is the audience? What is the purpose? What goals is the letter supposed to accomplish? The students are assigned a specific writer role: they are technical writers in an insurance company, whose task is to write, revise, and update the form letters that go out to policyholders. In this case, one particular letter (a policy cancellation letter) has caused some problems. The technical writer's job is (a) to assess the problem, and (b) to revise the letter to improve the situation.

Jim begins by assigning students to read the case outside of class. In class, Jim and the students discuss the project to determine "the facts of the case" in Sherlock-Holmesian fashion: Why is there a problem? Why does the company think the letter needs to be revised? Who is offended by the letter? Once the students are immersed in the case scenario, Jim assigns readings from the Documents and Principles sections of PWOnline. In Documents, the students read the material on the rhetoric of the letter. In Principles, they might read the sections on the purposes of professional writing and on ethics.

We do not always start with projects, and PWOnline accommodates other approaches to professional writing. One may begin, for example, by introducing a Document type or genre. Here is an example of how this might work.

Scenario: Teaching from a Document or Principles Oriented Approach

Joanne wants to introduce memos to the students in her business communication class. She asks them to read (before class) the discussions in the Documents section on "the rhetoric of the memo" and to look over several of the sample memos. She also assigns her students to read the section on evaluating writing (from Principles).

In class, Joanne starts by examining several of the sample memos, as well as several of her own samples she brings to class as handouts. In the discussion of each sample, Joanne asks her students to apply the principles of evaluation from the Principles discussion and to think about each sample in terms of how it accomplishes its aims as a memo, whether it is well organized and formatted for its intended audience and purpose, etc.

Joanne's teaching approach is to teach the genre of the memo by alternating between Documents and Principles: She and her students discuss samples, reflect on the principles that apply to that sample, then move to another sample. This kind of inductive approach helps her students learn how the genre works, learn how to evaluate good/bad instances of the genre, and build knowledge about how the memo works as a document.

Once the students have a grasp of the genre, Joanne moves to the Project section and assigns a memo project to her class.

Some Scenarios of Use In Computer Classrooms, Traditional Classrooms and Distance Education Courses

Scenario: Using PWOnline in the Distance Education Course

Amy teaches a distance education course in technical writing. Her syllabus for the course is on the web — and the schedule for the syllabus serves to set the agenda for each class period. Below is a segment of the class agenda section from her syllabus. (Please note that the links in her syllabus will NOT work! They are for illustrative purposes only.)

Notice that in each of the three classes in the table on the next page, Amy assigns the students readings from PWOnline. Some of the readings are from the theory overview section of PWOnline. Two of the hyperlinks take the students to project descriptions (their first two projects for the course). Because Amy uses a web syllabus, she can provide her students with live hyperlinks to the particular pages she wants them to read. Amy also includes in the syllabus links to her own syllabus, to her class notes, and to other instructional materials that she wants the students to have.

> **Note:** One issue that students raise in using PWOnline is the question of how large the reading assignment is. In the absence of page numbers, how do you know when you have read everything? On most of these web pages, there are multiple subsections and multiple links to other resources ... sometimes the students aren't sure what they need to read — so it's useful if teachers can provide the specific page URLs.

Figure 1-1: Sample Online Class Schedule

Technical Writing/Distance Education

English 234 Course Syllabus

Class Activity	Required Reading
topics • course overview • technology orientation • basic rhetorical principles of professional writing	**readings** • <u>ENGL 234 course syllabus and schedule</u> • <u>course policies</u> • ethics for professional communicators (handout) • <u>overview of rhetorical theory for professional communication</u> (PWOnline) **class notes** <u>234notes_5-16.doc</u>
topics • introduction to document revision project • basic theory of professional communication	**readings** • <u>description of document revision project</u> • <u>basic rhetorical theory for professional communication</u> (PWOnline) • <u>what is a project assessment memo</u> • <u>overview of rhetorical theory for professional communication</u> (PWOnline) **class notes** <u>234notes_5-24.doc</u>
topics • ethics • introduction to online documentation project	**readings before class** • <u>ethics discussion</u> (PWOnline) • ethics discussion (handout) **readings in class** • the class agenda (posted to the class list) • <u>project description: Online Documentation for Using MOOs</u> **class notes** <u>234notes_5-25.doc</u>

Scenario: Using PWOnline in Traditional Classrooms

PWOnline can work in the traditional classroom (that is, the classroom without computers). The disadvantage is that of course the teachers and students cannot "bring" the entire PWOnline into the classroom. However, students/teachers can print out selected material to bring into class (or use the print *Handbook* which provides copies of cases and some sample documents). Can teachers use an "absent textbook" in their classes? For teachers who are highly reliant on use of the textbook in class, PWOnline will probably not work that well. For teachers who use the textbook only occasionally or as a supplement to class activities and in-class presentations, PWOnline *will* work.

Scenario: Tailoring Projects for Local Situations

Steve is a technical writing teacher at a major private university. His class consists of junior and senior electrical and computer science engineering majors. Steve wants to use the Online Documentation project in his technical writing class — but he wants to make the project parameters more specific and localized for his own university. (He would like his students to test and revise the existing online documentation helping new students at the university activate their web accounts and hookup their dorm computers to the campus backbone. Of course, the case as written isn't that specific.)

Steve downloads the Online Documentation project — and revises the project by adding specific details and tailoring the assignment description for the context at his own University. He then uploads the project description onto his own web site and has the students access the case there. (He will to change the relative hyperlinks in the case to absolute links.)

Because Steve teaches in a traditional classroom, he makes print copies of the description page for in-class distribution to his students. This is not an ideal solution, as the teacher and students don't have in-class access to the support materials for the case (e.g., principles, templates). However, the case *is* usable in this fashion.

Scenario: Developing Distance Collaboration

Anne makes an arrangement with an instructor at another university to do the Corporate Web project collaboratively — as a way to give his students experience in doing online collaboration. Each student team has four members, two from each university. The four team members meet and collaborate on the project "virtually." They use e-mail to send messages and files to one another, and they use the PWOnline corporate web chat space for synchronous team meetings.

At several points during the project, Anne and the other instructor meet with all the students in both classes — in the chat space — to discuss how the project is going, to do project troubleshooting, and to reflect on the joys and challenges of distance collaboration.

Scenario: Using PWOnline to Support International Understanding

Gu uses PWOnline for his Introduction to Research class for technical communication majors. After completing the Job Banks project, he has the students internationalize their thinking by extending their job materials to search for work as trainers of teachers in another country. They research how the job letter and resume change in other countries, and then they chat with soem instrutors (and other visitors) Gu has found who have held jobs in other countries. Their experiences help them understand how employment documents are culturally-bound.

chapter 2

Site Map

PWOnline has four major divisions: Projects, Documents, Principles, and Resources.

Projects

You'll find writing activities -- from projects and cases to exercises -- in this section. What's the difference between cases and projects? A case usually provides a complete scenario for writers to work with (like the Big-1 case). It places the writer within a real, or realistic, scenario and asks the writer to play a given role within that scenario. Projects, on the other hand, may involve a client-based research simulation. This sort of simulation provides a general framework but asks writers to build their own cases within that framework. For instance, for the Corporate Web Project, the simulation part is that writers are working for a consulting company, Web Page Solutions. They are then expected to develop their own clients and projects, real ones ideally, within the parameters of the simulation. Exercises are intended to give you experience applying the principles that you'll use to develop projects and cases. Chapter 3 of this *Handbook* includes more information on the Projects component of PWOnline.

Documents

You'll find samples of writing (produced by students as well as professionals) and templates organized by genre, for example Memos or Employment Documents, in this section of PWOnline. You'll also find pointers to general discussions of principles and projects that lead to production of each type of document. You'll find more information about the Documents component of PWOnline in Chapter 4 of this *Handbook*.

Principles

You'll find explanations of the principles of professional writing in this section of PWOnline. Principles includes textbook-like discussions of writing in organizations, writing particular to specific types of documents, and completing writing-related tasks such as page design, field research, and editing. You'll find a brief overview of the Principles section in Chapter 5 of this *Handbook*.

Resources

You'll find pointers to a wide array of web-based resources in this section of PWOnline. The Resources section functions in much the same way a bibliography might function in a more traditional textbook. You'll find an abundance of pointers to sources on the Web that will assist you in locating information, developing Web materials and print documents, and becoming a professional writer. You'll find a brief overview of the many resources available in the Resources section in Chapter 6 of this *Handbook*.

The contents of each section are listed in the table on the following page and are described in further detail in the next four chapters of this Handbook, and the PWOnline site itself has many cross-references. But, everything you need -- as usually is the case on the Web -- is not handily stored in one place. You need to browse around. Of course, should you need quick access, you can use the site's search engine (it is prominently displayed on each page).

Table 2-1: Professional Writing Online Sitemap

Projects	Documents
• Software Learning Initiative • Documentation • Evaluating the Design of Online Documentation • Analysis of Writing Practices • Analyzing Professional Contexts • Allied Mutual Insurance Case • Airbag Case • Copyshop • United Drill • The Corporate Web • The Big 1 Case • K-Mart Case • Employment • Lester Crane Case • Technology Access • NCAA Bat Standards • Software Bug Case • Job Banks • National Electric • E-Commerce • Feedback	• Introduction • Memos and E-Mail • Letters • Reports • Policy, Manuals, Handbooks • Employment Documents • Promotional Materials • Instructional Documents • Oral Presentations • Feedback
Principles	**Resources**
• Introduction • Overview • Understanding Readers • Social and Cultural Issues • Shaping Texts • Analyzing Workplace Writing • Building Arguments • Managing Projects • Arranging Information • Research • Writing Reports • Writing Online • Usability Testing • Style • Document Principles • Feedback	• Introduction • Instructor Resources • Search Engines • Searching for Information • Corporate Web Sites • Web Design Resources • Graphics and Visuals • Online Documentation • Internet Use Statistics • Usability Resources • Design Tips and Templates • Professional Writing • Citing Sources • Job and Career Resources • Professional Writing Programs • Feedback

chapter 3

Projects

The Projects section of *Professional Writing Online* provides real (or, at least, realistic) scenarios for projects in professional writing. The projects listed below include both cases and client-based research simulations. Variations on these projects are of course possible, and users are encouraged to download and customize these projects for their own uses.

Table 3-1: Projects in Professional Writing Online

Project	Topic(s)	Use(s)
key to symbols	▼ could be used early in a course ● well suited to business communication course ■ well suited to technical communication course	
Software Learning Initiative	computer documentation (quick-reference sheet) and technical training (live tutorial). (three-day project)	an early project to learn software for computer classroom
Allied Mutual Insurance Case ▼●	letter writing, corporate image, handling customer complaints	a case that addresses complaint letters
Technology Access ▼	distance education, computer technology, short report/memo	an early project for a distance education class
Big 1 ▼●	corporate relations, bad news	case that ghost-writes a complaint letter
United Drill ▼	policy writing, sexual harassment (two-week project)	case that practices research and policy-making
E-commerce	electronic commerce, business feasibility, web business, research report	project researching the web

Table 3-1: Projects in Professional Writing Online *(continued)*

Project	Topic(s)	Use(s)
K-mart ●	web site management, corporate image, personnel	case that generates a bad news letter
Copyshop	legal issues (copyright), policy writing	case that practices research and policy writing
Airbag Case	instruction writing, safety and accuracy	case that revises a complaint response letter
Software Bug Case ■	translating technical data for cutomers, writing bad news letter, software liability, corporate image	case generating letter explaining policy
NCAA Bat Standards	sports, developing templates for letters, online research, explaining technical information (one-week project)	case generating template letters
Lester Crane ●	international communication (Middle East), scheduling, client relations, politics	case that practices writing reports to internal and external audiences
Corporate Web	web page development, usability, corporate image	project or simulation that serves as major group project
Job Banks	evaluation of online job banks for various professions (two- to six-week project)	project that analyzes web resources and may include workshop
Documentation ■	online computer documentation (tutorial), usability, instructional writing (four- to six-week project)	real project for online documentation
Editing Online Documentation ■	online computer documentation, instructional writing (one-day project)	exercise practicing peer review; works well with documentation project
Analysis of Writing Practices	investigating and theorizing about the ways writing functions in a particular field	genre project that gathers and evaluates writing inside students' fields
Analyzing Professional Contexts	conducting and analyzing field research about your professional context to become a better communicator within that context	field research project that investigates practices in a professional organization
National Electric	short case on creating visuals	exercise constructing memos that include visuals
Employment Project	job application letter and resume (one- to two-week project)	project that has students apply for a real jo

Software Learning Initiative
by Johndan Johnson-Eilola, Clarkson University

Context

As your company begins work on a massive new initiative, it becomes evident that many people — including yourself — will need to come up to speed on a range of new programs. In order to help bring everyone up to speed quickly, the project leader, Indra Samuelson, announces the "Software Learning Initiative." Briefly put, this means that each of the project teams will be responsible for training the other teams to use one piece of software or set of programming skills. The training will include both a half-hour live tutorial run by your team as well as a one-sheet quick-reference card for users.

Deliverables

For this short project, your team will need to complete three discrete deliverables:

1. **Informal Proposal** (via email to all teams)

   Your team will begin by meeting with all project teams to discuss possible programs to cover in each tutorial. The team leader may insist on some of the programs to include, but you will want to be ready to suggest others. (When deciding what to cover, think carefully about the broader project your teams are working on.)

   After the meeting, your team will send an email message to the project-wide email list specifying exactly what program, language, or skill your team will cover and why other teams should be interested in attending (and paying attention to) this tutorial.

2. **Tutorial** (script and actual performance)

   Your team will also need to develop a script outlining exactly what you'll be presenting to the other teams during the tutorial. The script should include not only what the presenter from your team will say, but also what will be shown on the overhead projector to your users, checkpoints for evaluating whether or not your audience is keeping up, and other notes.

   The first draft of the script and the final draft must both be submitted to your advisor Ms. Samuelson for her records. She'll be commenting on the first draft to help you make sure you're covering material adequately and in a way that your audience will understand and appreciate.

3. **1-Sheet Quick Reference Card**

    In order to help the other teams remember and apply what your team teaches them, you'll also need to design a one-sheet quick-reference card for them to take away. The card must be carefully designed to fit a maximum amount of useful information into a small space. You'll need to think about where and how your audience works in order to decide what size and format to use (you may want to use some other size besides US Standard Letter, 8.5 x 11 paper in portrait orientation).

    You'll also need to submit both the draft and the final to Ms. Samuelson for her comments and her records.

Discussion Questions: Software Learning Initiative

1. **Brainstorming Key Tasks.** It may seem that 30 minutes allows you enough time to do whatever you need. But, unless you engage your audience in information and tasks that are relevant to their needs, you will find the session difficult. User and task analysis are your allies in developing an effective presentation. What are the key tasks for your users?

2. **One-Sheet Documentation.** Have you used a command card or other one sheet reference document that you found helpful? What was on that card that helped you use the software? What seemed odd or off-target? Can you think of some tips that will ensure your group's success? [Note: If you haven't used this type of document successfully, try to think through why not.]

Designing Online Documentation
by James Porter, Case Western Reserve University

Context

Your assignment is to create documentation for use at your company or university. For example, you could write a tutorial for users new to using Netscape Composer as a tool for web authoring. Or you could teach users how to access and interact in a MOO server.

To give a more specific example, students working at Case Western Reserve University could write documentation to support the use of Case Western Reserve University's new i-drive interface, introduced for use in Fall 1999. (Note: "i-drive" is both a server and an interface. Every student and teacher at CWRU is given 25 MB of space on i-drive for uses related to instruction — see <http://idrive.cwru.edu/>. Students and instructors can use the space to distribute or collect files or to serve web pages. There is a standard home page interface that functions as the starting point for users — for a sample, see <http://idrive.cwru.edu/jep21>.)

The project as written below is directed toward the last example above. As it is written, it could only be completed by those who have accounts on Case Western Reserve University's i-drive system — that is, CWRU teachers and students. However, those working elsewhere can do the same sort of project and apply the same principles and procedures to create documentation for software applications or systems at their own institution or company.

Deliverables

All or some of the following documents might be produced in conjunction with this project:

- project plan for participants (including Gantt chart) — see sample project plan in PWO
- interview questions for users
- usability test procedure
- usability test report (results collected and analyzed, recommendations)

- documentation (beta version) documentation (revised version)
- project assessment
- team evaluation

Discussion Questions: Designing Online Information

1. **Defining Users?** Who are the users of i-drive? Anybody with Internet access can login to i-drive as a guest and access shareable files. However, the documentation you will produce is intended for those with i-drive accounts — CWRU students and teachers who need to learn how to setup i-drive for their own classes and other instructional and academic purposes. Further, for this project you should specify a particular subgroup: CWRU *writing* teachers and students in writing classes.

   But this raises an important issue: How specifically should you define your user group? Should you specify even further — say, to CWRU students and teachers in a particular course (e.g., English 398, Professional Communication for Engineers)? If you define your user group too broadly, your documentation is perhaps not all that useful for different users. If you define your user group too narrowly, the documentation is of limited use. What makes sense to do?

2. **Print vs. Online Documentation?** Should your documentation be print or online — or usable in both formats? Obviously any piece of online documentation could also be printed out, but some documentation is *designed* for online use and some is not. Your i-drive documentation will be stored online and it can always be printed out — but you have to determine how it should be *designed*. How are users likely to use the documentation? (For instance, web-based documentation is fairly useless if the intended users are not already on the web. Some online documentation is rendered useless when the users can't keep the documentation open *and* perform the tasks simultaneously.)

Notes

1. **Approach to design.** Apply a "user-centered approach" to your documentation design (see Johnson 1998). Briefly, this means that you should focus not primarily on the design of i-drive, but on *the needs of the user*. What do users need to know how to do? What work will they be doing? What tasks would they like i-drive to support? The documentation should be oriented to their needs.

Approach to Design	Example #1 (from Redish, Felker & Rose, 1981)	Example #2
a **systems-oriented** approach (focuses on system design)	"All station logs which are required under those provisions of this part pertaining to the particular classes of stations subject to this part shall be retained by the licensee for a period of one year from date of entry and for such additioanl period as required by the following subparagraphs."	"I-drive is divided into five places for the storage of information: The "Share" folder is the location for word-readable files."
a **user-centered** approach (takes the perspective of the users)	"You must keep a radio log...You must keep your radio log for at least one year after the day of the last entry in the log."	"to turn your homework assignment eletronically, login to your instructor's i-drive home page and look for the Dropbox."

Applying a user-centered approach means involving actual users at all stages of the design process.

Project Stage	Approach to involving users
Early -- interviews, observation, focus groups	Start by talking to users. Interview them to find out what they know about i-drive, what they'd like it to be able to do. If they already have i-drive accounts, observe them using those accounts to see what they are missing. If you can, collect a small grou of users together for a focus group discussion (which can often generate more and different information than simple one-to-one interviews.)
Middle -- usability testing and expert review	Once you have developed a prototype (beta version) of your documentation, test it with actual users in a usability test session. Be sure to test the documentation with multiple users and with a range of user types (male/female, low-end users/high-end users, novices/experts). Also, test your documentation with expert reviewers -- for example, you might test the documentation for technical accuracy by giving it to a member of the i-drive development team.
End -- ongoing evaluation and revision	Based on the findings from your usability sessions, revise and then publish your documentation. Be sure to provide a feedback mechanism in it (a phone number or a mailto: link) that allows users to give you information over time as they use it. As i-drive changes or as you learn more from users, update the documentation.

2. Collaboration and project planning

   In addition to collaboration with users, you will produce this documentation in groups of 3-5

   - **Step 1** in working as a group is to meet and determine a plan for doing the project. A plan starts with the group setting goals — both in terms of products to be produced ("deliverables") as well as outcomes: what do you want to have happen as a result of your work? what products will enable you to meet those goals? what measures of success will you apply? how will you determine whether your project meets the needs of users?

   - **Step 2** is to define who your users are — not always an easy or obvious task. Can you write this documentation for teachers and students simultaneously? Or does each group need its own documentation? There may be many needy users out there — but pragmatically, can you write documentation to suit all their needs?

   - **Step 3** is to build a production schedule: what deliverables are due when? Some due dates are *fixed* (the ultimate due date for the assignment). Other due dates are flexible, depending on the group's activities, when you can meet, who can do what. Step #4 is to coordinate tasks and responsibilities among yourselves. What tasks should be done together as a group? What tasks should be divided up and done individually? Steps #3 and #4 should probably happen together; the Gantt chart is the tool you use to coordinate project tasks with your schedule.

   - **Step 4** is to coordinate tasks and responsibilities among yourselves. What tasks should be done together as a group? What tasks should be divided up and done individually? Steps #3 and #4 should probably happen together; the Gantt chart is the tool you use to coordinate project tasks with your schedule.

   Intelligent and careful planning is an incredibly important facet of this project. Be sure to build some "reality time" into your schedule — that is, to allow for project holdups, sicknesses, missed deadlines, and the sorts of glitches that inevitably happen in project life. Work up this planning into a formal report, called a project plan.

Evaluating the Design of Online Documentation
by James Porter, Case Western Reserve University

Context

Your assignment is (a) to evaluate the design of a sample piece of online documentation, and (b) to make recommendations for improvements in the design. (Check the list of Resources Outside PWOnline for samples.)

Your role in this exercise is as an editorial advisor providing expert design feedback. Note that you are not being asked to evaluate the usability of the documentation (i.e., whether or not real users can use it to complete a task). You are not being asked to revise the documentation yourself. Rather, you are evaluating its adherence to recommended principles and practice for online text and you are giving advice to the writer(s).

Your audience for the project is the writer (or writers) of the documentation. Your evaluation and recommendations should, thus, be couched in positive terms, not discouraging ones. Encourage the writer and make positive recommendations for change. (Don't slam the writer's work.)

Deliverables

A memo to the writer of the documentation which (a) evaluates the design of the documentation, and (b) recommends changes to improve the documentation (cc: your teacher).

Discussion Questions: Evaluating Online Design

1. Should design principles be adapted for different circumstances — different audiences, different genres, different purposes? More specifically, how do the principles for online information change:

   - when the online information is documentation or instructional text (vs. other forms of online information)?
   - when the users are novices (vs. intermediates or experts)?
   - when the online information is reference documentation vs. tutorial documentation?
   - when the information is intended to be printed out (vs. used/ read online)?

2. Where do design principles end and other sorts of principles begin? What will you do in this exercise if you run across obvious technical errors in the documentation? Should you comment on the context and usability of the documentation, even if that is not your assigned task? What is the extent of your role as editorial advisor?

Notes

1. **Developing your expertise.** *The first step in preparing yourself to do this exercise is to make sure that you are acquainted with the principles for effective design of online documents. Make sure to read the relevant sections in PWOnline (noted above) and to check the Yale/CAIM guide to web page design.*

2. **How much commentary?** Your role is to provide overall editorial advice — but not to actually correct or redesign the documentation. Decide what is most important to focus on. Highlight the main principles — and refer to specific places in the documentation where improvement could be made. Do not do line-by-line editorial commentary or interlinear revisions.

Analysis of Writing Practices
by Johndan Johnson-Eilola, Clarkson University, and Michele Simmons, Miami University of Ohio

Context

You will gather and analyze a professionally written document from your field in order to investigate and theorize about the ways writing functions in your specific writing context, community, and field. You may choose any text that allows you to examine the type of writing and writing practices that occur in your field. Examples vary across fields, but may include:

- technical reports
- progress reports
- project proposals
- performance or safety evaluations
- programmer's manuals
- memorandums
- textbook excerpts of managerial or organizational strategies
- journal articles

Deliverables

All or some of the following documents might be produced in conjunction with this project:

- **Document analysis.** Using the two resource links for reference, and the provided questions as a guide, you will construct a two page analysis of how writing is produced and works in a particular context within your field. The analysis should be in memo format, with an introductory paragraph and a heading for each of the five categories of questions included below. Address the memo to your instructor.

You can download a Word 6 memo template to use as a starting point for formatting your memo.

- **Written document from your chosen field.** Attach a copy of the document you analyzed to your memo.

Document analysis

Begin by reading through the five categories of questions below: purpose, writing roles, types of arguments, structure, and function. Then read through the information in the two links listed in the resource section, *Analyzing Workplace Writing Situations*, and *Building Arguments*, which will address each category. After reading the resource materials, use the questions to construct your meta-analysis.

Purpose

For what purpose or purposes is your document written? Be sure to indicate the context and the primary audience for which this document was written, as well as whom the document is likely to affect.

1. **Writing roles.** Which of the following roles could be played by this writing?

   - teaching
   - selling
   - advocating
   - judging
   - researching (or reporting research)
   - other? Specify.

   Explain how this role is used to support the purpose of your document.

2. **Types of arguments.** What *type* of argument (e.g., policy, fact) is being made? Give a few examples of how this type of argument is made and indicate the page and paragraph number. What is the overall argument of the document? Explain one other way this argument could be made.

3. **Structure.** How is the information structured? Pick out three

strategies your document uses to manage information (e.g., time, space, hierarchy, alternatives, steps, topics, matrices) and indicate the page and paragraph number where this strategy occurs.

4 **Function.** Common uses of professional/technical documents include: to do, to inform, to learn, to choose, to archive.

- For what function was your document written? Certain structuring strategies are more useful for certain types of documents.
- Does your document use appropriate structures for the function of the document? Give an example.
- Explain one other structure strategy that could be used appropriately to manage information in your document.

Analyzing Professional Contexts
by Amy C. Kimme Hea and Melinda Turnley, Purdue University

During the Analyzing Professional Contexts project, you will conduct and analyze field research of your professional context. This field research will consist of an interview, document analyses, and an observation. After coding and analyzing your field research data, you will plan and draft a report discussing significant issues related to your field.

This long term, multicompenent project will introduce you to data collection methods and project planning and development. These processes will be beneficial in relationship to not only to other course projects but also writing projects that you will be assigned as a professional in the workplace. This project was developed from the "Analysis of Writing Practices" project that is also available in PWO.

What is a "context"?

A context is a dynamic set of relationships among people, institutions, documents, technologies, etc. These relationships inform and are informed by certain standards, judgments, beliefs, assumptions, and values. Becoming a member of a professional context is a process that requires you to invest in and understand these relationships and the various ways in which they function. To research your own professional context, you will collect data from academic and other workplace settings in your major or specialization. The purpose of this field research is for you to investigate directly specific people, sites, events, situations, and documents that are part of your professional context. This investigation will require you to consider how these specific examples relate to issues of:

- power/authority
- knowledge/expertise
- status
- worker-worker or student-student relationships
- management-worker or teacher-student relationships
- initiation of contact or discourse
- completion of contact or discourse

Why research your professional context?

If your supervisor requests that you write a report outlining your recent work on a project, you will need to determine what the standards for effective reports are for your context. To better understand such standards, you likely will speak to colleagues, collect past reports and related documents, and make note of procedures and precedents related to the creation, distribution, and evaluation of written products within your work site.

The field research methods that you will use in this project provide general strategies for investigating rhetorical expectations in your field. Thus, although you may not always be granted the time and opportunity to formally conduct interviews, document analyses, and observation, these field research methods provide starting points for the planning and drafting of professional documents.

For this project, your research report will provide you an opportunity to reflect upon the ways in which these research strategies provide insight into issues that shape the standards in your field. Such issues are important to any field, but your Contextual Analysis Report will highlight the ways in which you became more aware of certain issues during your own research.

Project benefits
Carefully considering relationships and issues within your professional context can help you to become a better writer. For you to write persuasive and effective documents in your field, you must learn to assess the potential audiences and purposes for those documents. Throughout this project, you will hone your assessment skills through research, analysis, and drafting.

Your general project goals are to:

- Become more aware of writing as a social act.
- Understand writing as a set of contextualized process.
- Relate writing strategies to workplace contexts.
- Conduct and manage a complex writing/research project.
- Create a persuasive report based upon research.

Deliverables
All or some of the following documents might be produced in conjunction with this project:

- **Interview transcript.** Conduct at least one interview of an established member of your professional context (e.g., a professor in your field, a senior employee in your workplace, a researcher, etc.). After conducting your interview, you will submit a typed transcript of your conversation, including both your questions and your interviewee's responses. Consult additional interview information in order to plan and conduct your interview. Use the downloadable Word form to complete your interview transcript.

- **Document analyses.** Collect and analyze at least 2 documents from your professional context (e.g., policy statements, curriculum guidelines, style manuals, journal articles, documentation, memos, web sites, reports, forms, etc.). After selecting and analyzing your 2 documents, you will submit both copies of the documents and your completed analysis of them. Consult additional document analysis information in order to plan and compose your document analyses. Use the downloadable Word form to complete your document analyses.

- **Field notes transcript.** Observe and compose field notes for at least one site within your professional context (e.g., classroom, meeting, conference, lab, industrial site, office site, etc.). After selecting and observing your site, you will submit your field note transcript of your observation. Consult additional observation information in order to plan and conduct your observation. Use the downloadable Word form to complete your field notes transcript.

- **Data coding grid.** After conducting all of your field research, sort and analyze your data. You will read, mark, and code your interview transcript, document analyses, and observation transcript in order to complete your data coding grid and identify which issues for analysis you want to highlight in your plan and report. Consult additional data coding information in order to code your data and complete your data coding grid. Use the downloadable Word form to complete your data coding grid.

- **Contextual analysis plan** . After completing your data coding, complete a contextual analysis plan that will function as an outline for your contextual analysis report. Consult additional contextual analysis information in order to plan and draft your contextual analysis report. Use the downloadable Word form to complete your plan.

- **Contextual analysis report** . After completing your contextual analysis plan, draft your contextual analysis report. Consult the contextual analysis information in order to plan and draft your contextual analysis report.

> ### Discussion questions: Analyzing Professional Contexts
> For this project, you should consider your professional context. The following prompt is developed to guide you in starting this project:
>
> 1. What is your professional context (i.e. your field, major, or specialization)?
>
> 2. Provide examples of elements that make up your professional context.
>
> Use the downloadable Word form to complete your professional context assignment.

Customer Complaints at Allied Mutual

Reprinted with the permission of Scott Jones, Cornell University, Copyright 1994

Allied Mutual Insurance Company sends form letters to clients notifying them when their policy has been cancelled. The letters are in a computer, which prints out letters with appropriate names and cancellation reasons as required. Recently, agents who sell Allied Mutual Insurance have reported that clients are angrily complaining about the tone of the cancellation letters.

Context

Allied Mutual Insurance Company is a member of a group of associated insurance companies known as Aligned Insurance. Allied Mutual offers auto and property insurance policies, while the affiliated companies offer life, health, and other policies.

Allied Mutual, and its affiliated companies, place a great emphasis on treating their customers well. They see the companies' main goal as serving their clients by being there to help them at times of great loss. As such, they always strive to communicate honestly and clearly with policyholders. This is particularly important to the CEO of Allied Mutual.

Insurance policies and termination
The letter in this case is sent out once a policy has lapsed ("run out")

because the company has received no payment. The company continues a client's policy for three weeks after the payment was due, partially as a courtesy, and partially to give the client time to purchase other insurance. After this point, the policy is terminated.

If a client pays the policy after receiving this letter, the policy may be continued. The company reviews the policy again, just as if the client were a new customer, and decides if the customer should be retained. New clients must meet higher standards than existing clients, so it's possible that after reviewing the client, the company may choose to no longer cover the client.

In short, the client's policy is going to run out. If they pay the policy, the company may continue it or they may refuse to renew the policy and return the payment

Problem
The company's agents are upset. They maintain that clients are unhappy with the current cancellation letters. This is important to the company for a variety of reasons:

- The company does not want bad word of mouth from former clients.
- Misunderstandings may have occurred, such as payments getting lost in the mail, and polices have been cancelled, incorrectly.
- Clients may have mailed payment after the due date, but before this letter was sent.
- Clients may have been customers for a long time, and expect to be treated as especially valued.
- Clients may hold other policies with the company, such as home or life, and they may cancel those if the letter angers them.
- Agents are very important to the company, and must be kept happy. The company has been having some difficulty retaining agents, so this is an important issue. If agents think the letters are bad, then they must be revised.
- None of the agents have reported any specific problems with the letters — just that clients are unhappy with the letters.

Deliverables
You work for Allied Mutual and have been asked to rewrite one of the

policy termination letters (the one for a lapsed policy) so customers will not become angry, upset, or offended while reading it. No one at the company is quite sure why clients are becoming so upset with the current letters, so you must determine this on your own.

The letter you will revise, a form letter for Kansas, is available via this link: *Lapsed Policy Letter*. You should keep several things in mind as you work to revise this document, including the following:

- the document should look professional
- clients must be treated with respect, not as "deadbeats"
- it must be made clear that their policy has been terminated, why it has been terminated, and exactly when coverage ends
- it must be made clear that Kansas law requires them to have auto insurance
- clients should understand that if they pay the policy it *may* be reinstated at the company's discretion
- this is a legal document, so all wording is very important. However, the legal department will review the letter before it is finalized, so while you should be concerned with trying to word the letter in a legally acceptable manner, you are not ultimately responsible for ensuring its legality
- the letter will also be reviews by your boss, your boss's boss, and the CEO of the company
- while you want to maintain a good relationship with the client, you do not necessarily want to continue the policy
- Also, please note that the person to contact listed on the letter is the client's agent. The agent resides in the same region as the client.

Discussion Questions: Customer Complaints at Allied Mutual

1. **Letter problems.** Why are clients disturbed by this letter sent to people who have been irresponsible in their payments?

   Begin to answer this by thinking about what a good business letter accomplishes (e.g., explain a position, get reader to act, gain agreement, build trust, defend one's actions, and so on). After reading it from the point of view of accomplishment, what might the <u>lapsed policy letter</u> Allied Mutual has been sending accomplish? How does it seem to treat its reader? Consider, too, the tone built by a good letter (e.g., cooperative, intending to inform/explain, friendly if businesslike tone, and so on). What tone does this letter build? If you received it, what would you think that company thinks of you and the businsse you bring them? How does this letter shape the receipient's relationship with her/his agent and with Allied Mutual?

2. **Bad responses.** What kinds of responses would be inappropriate to add to your revision? Think here about information, about tone, about instructions you give to the letter's recipients, and so on.

3. **Technology concerns.** What production concerns are present? For example, if the letter is produced at the home office in Kansas City (as it presently is) from database information (as it presently is) how can you develop a response that is personal and yet usable in more regions than just Lafayette? What tensions are present in the struggle between the production requirements of a form letter and the agent's desire to personalize where possible (because the letter is important to shaping relationships)?

Notes

The situation in this case, the task itself, and the accompanying letter were all taken from real life. Every element of this case is exactly as it was in real life, except names and places have been changed.

Airbag Safety and Customer Service at Vanguard Motor Company

Reprinted with the permission of Bill Hart-Davidson, Rensselear Polytechnic Institute

Context

In the middle of your second week as an intern at Vanguard Motor Company, you get a note from your supervisor saying that she wants to speak to you about a new project. Your work, to date, has been interesting, if slightly mundane. You are mostly revising procedure manuals according to updated company standards and then checking these against existing state and federal regulatory standards.

In the note, your supervisor mentions that it's your "excellent writing skills" that are in demand for this new project.

Customer concerns over airbag safety

Your supervisor, Barbara Jones, explains that Vanguard has had a great deal of mail coming in relating to airbag safety and the question of how to get an on/off switch. To date, management feels as though they have handled these customer complaints and inquiries ineffectively — mostly on a case-by-case basis and without much consistency across cases. What they'd like you to do is to draft a short letter that could be sent to customers who inquire about the safety of their airbag and/or the possibility of getting an on/off switch installed in their automobile.

Ms. Jones also explains that you won't be starting this project with a blank slate, as several recent incidents have sparked management's attention to the need for a better response to customer concerns. In particular, your boss refers to a customer who wrote a letter inquiring about an on/off switch and who identified herself as "small boned" and, therefore, at risk of injury from the airbag installed in her automobile. The response letter she received from Vanguard was, in Barbara's words "defensive and, well, rude."

Moreover, the response letter offered no instructions about how to properly use the restraint equipment in the vehicle to avoid injury - hence the need for a "tech person" to write this letter.

Deliverables

It becomes clear to you that your task here is not to build a letter from the ground up, but to revise the letter sent to the "small boned" customer so that it can be sent to other concerned customers. It seems to you that the letter should:

- Offer instructions about how to properly use the restraint equipment to avoid injury

- Be in step with recent federal regulations regarding the installation of on/off switches in airbags (the original response letter was written before the regulations were finalized, which is one reason the letter sounds so vague and stand-offish.

- Accurately represent Vanguard's genuine concerns about customer satisfaction and safety as well as their commitment to standing behind the quality of their products

Ms. Jones also informs you that the letter should be short — no more than a page for the text of the letter and a single page of instructions — and doesn't have to go into detail about the federal regulations concerning on/off switches. You're happy to hear that because the deadline — one week away — doesn't give you much time to do research.

Complication: The national debate over airbag safety

The National Highway Traffic Safety Administration (NHTSA), a division of the Department of Transportation (DOT) has recently passed new guidelines permitting the installation of airbag on/off switches in automobiles of customers who meet specific risk-factor requirements. On/off switches on Vanguard vehicles are retrofit and are not available as a general option, due to the NHTSA ruling. The attached Netscape file contains some links to NHTSA information and a few other relevant sources which you may find useful.

Because on/off switches are available only to those persons who are in designated risk groups, the national trend has turned toward educating the public about ways to avoid injury through proper use of restraint systems.

> **Discussion Questions: Airbag Safety**
>
> 1. Examine several of the websites that dispense information about airbag safety. Take the point of view of the consumer protectionists. What are the issues in airbag safety for consumers? How do the car manufacturers' pages talk about airbag safety: do they address the consumer concerns, or do they take up different issues, or do they ignore airbags in their websites?
>
> 2. Do you have experiences with airbags deploying? Why did it happen? Did it contribute to passenger safety or did it have some other effect? Based on your own experiences, what attitude to airbags to you take to this project?

Copyshop Copying Policy Case

by James Porter, Case Western Reserve University

Context

You have just recently been hired as the new assistant manager of Copyshop, a local and independently owned copying business located near a large state university. One of the primary services Copyshop provides is to create "coursepacks" for instructors at the university. (Teachers collect print materials they wish to distribute to their classes, bring these materials to Copyshop, and Copyshop reproduces them in a bound format called a "coursepack." Sometimes these coursepacks consist entirely of the teacher's own material, but usually coursepacks consist of material collected from a variety of print sources, such as book chapters, newspaper articles, journal articles, and printed out electronic documents. Often these collected materials are copyrighted.)

Currently the Copyshop does not have a written policy regarding the creation of coursepacks. But with the controversy created by recent law suits involving copy shops (e.g., the Kinko's case and the Michigan Document Services case), the manager and owner of Copyshop, Don

Magnum, is becoming increasingly worried that Copyshop might be held legally liable for copyright violations in coursepacks. Don feels that Copyshop needs a written policy specifically addressing coursepacks.

Don asks you to do some research on copyright issues involving coursepacks and to develop a copying policy for Copyshop. He asks you to write a report in which you present your findings, append a draft of the copying policy, and provide a rationale for the policy. The policy should be clear and direct — but it should also cover the diverse materials instructors include in their coursepacks.

Audiences

This policy you develop will serve three purposes for three different audiences:

1. For Copyshop *employees*, the policy should serve an informational purpose, telling them whether they need to seek permission for material instructors bring to Copyshop.

2. For *university teachers*, the policy should serve an advisory purpose, letting them know what materials require copyright permission (and, thus, guiding their choice of materials for compilation).

3. For *lawyers* and *publishers*, the policy should serve the function of letting them know that Copyshop does respect copyright laws and does have a clear and legal policy for seeking permissions to reproduce copyrighted materials. (This third purpose is also a protective function: if Copyshop is ever sued for copyright violations, the policy should indicate clearly that Copyshop does have a legal copying policy in place.)

Deliverables

The following documents might be produced as part of this project:

1. a policy statement, indicating Copyshop's position on coursepacks
2. a short report written to Don Magnum — the report should (a) present information about recent copyright cases involving coursepacks, and (b) provide a rationale for your recommended copyright policy

> ### Discussion Questions: Copyshop Copy Policy Case
>
> Your report should answer several questions Don has about coursepack copying policy:
>
> 1. For what sorts of materials do we need to seek copyright permissions? Does our policy need to address the issue of copying paper versions of electronic materials (e.g., web pages, e-mail)? Are electronic documents subject to the same copyright restrictions as print documents? What about copying graphics?
>
> 2. Do we need to seek permission every time a teacher makes any changes in a coursepack?
>
> 3. Can we reproduce entire books or textbooks at all?
>
> 4. What procedures are involved in getting copyright permissions for coursepack material?
>
> 5. Who is legally liable for copyright violations in a coursepack? Is the instructor at faul — or does the copy shop take the blame? Don asks, How much trouble can I get in if I am sued for a coursepack copyright violation?
>
> 6. Do we need to pay for an intellectual property lawyer to review our policy?

United Drill Case

by Patricia Sullivan, Purdue University

Context

Your boss Pat Chalmers, a department head at United Drill, Inc., is supposed to review a draft of a harassment policy for your division (North American Operations). United Drill is an internationally successful and conservative midwestern manufacturer of custom parts with ~ 450 employees in 12 countries and a workforce that has 15% women, 18% ethnic minorities and little sympathy for nonChristians or those living alternative lifestyles. Melvin White, the Vice President in charge of the division (and Chalmers' boss) has determined that such a policy will impress the EEOC (Equal Employment Opportunity Commission) as they investigate the company because of a charge brought by an employee who claimed to be terminated in retaliation for her charge of sexual harassment.

Because of problems in several of your department's projects, Chalmers, who is not in favor of such a policy, has not had time to research a plan of attack to thwart the policy. To make a long story short, you are given a draft of the policy and then must look into the matter and assemble some strategies and data for Chalmers to use in the policy review team meeting.

Chalmers wants to use your analytic memo as the basis for a memo to the other members of the review team.

Deliverables

You need to complete two documents for this project:

1. **Problem-Analysis Memo.** As you start the project, Chalmers asks you to write up a brief problem-analysis memo, in which you specify your understanding of the problem, your plan of research, and any questions you have. This will help to make sure you and Chalmers agree on the problem before you get too far in your research.

2. **Policy Review Memo.** The main memo you'll write will be used by Chalmers in the Policy Review Meeting. The memo should have arguments backed by data that challenges the advisability of adopting a formal harassment policy. It is important in this memo to remember that White is in favor of the policy. But, Chalmers adds, White's reason for endorsing the idea may be related to the EEOC investigation. Your task is to gather some data on bias claims and to figure out the ways in which the unit can respond.

> ### Discussion Questions: United Drill Case
>
> 1. Who is Pat Chalmers, and why has he/she asked you to write this memo on sexual harassment?
> 2. How well does the proposed policy meet the standards of federal laws dealing with discrimination and harassment?
> 3. How have other comparable companies been affected by discrimination and harassment problems?
> 4. How would the proposed policy affect **you** as an employee working at United Drill? How does it protect, or constrain, your ability to do your job?

The Corporate Web Project
by James Porter, Case Western Reserve University

Context

The Corporate Web Project is a major, client-based research simulation that takes writers through the process of conducting research for a company or organization and recommending (and, in some cases, actually implementing) changes based on that research. As the title suggests, this project has to do with corporate web sites. Writers will work for a simulated company (the professional writing class?) called Web Page Solutions, Inc., or WPS. This company exists to advise other companies, its clients, about the design and implementation of web sites: Should a company have a web site? (Is it worth the investment of time and personnel?) Is an existing web site meeting users' needs? Should a company upgrade, revise, or redesign its web site? Chiefly, WPS employees work as researchers and writers (and occasionally as web designers). They investigate their clients' needs, do appropriate research, and produce reports to help clients make decisions about their uses of the web.

Many variations of this project are possible, but the writer's primary task as a WPS employee is to locate a business client who needs information about web sites. Writers might help a client decide whether to develop a corporate web site (if they don't already have one), or whether to upgrade/revise an existing site. Writers could do a usability analysis or evaluation of an existing site. Once they have performed the initial study, writers could follow up their recommendations by actually designing or upgrading a web site for their client. (Note: Business writers are more likely to do feasibility studies and evaluation reports, based on a business or market analysis. Technical writers and web authors are more likely to do site usability studies and to actually implement web design changes.) This project can be done individually or collaboratively, but it is designed primarily as a collaborative assignment and it is designed primarily to produce a major recommendation report. (You might look at the sample report included in PWOnline: Recommendation Report: Design and Benefits of a Web Site for the Revegetation and Wildlife Management Center.)

For additional notes about context for the Corporate Web Project, see *The Corporate Web Project: Understanding Context* in PWOnline.

Deliverables

Depending on how the project is designed, writers could produce — as individuals or in teams — many different types of reports and documents for this project, including:

- initial project proposal
- project planning report for team writing project
- evaluation or usability report (analyzing and evaluating the current web site)
- progress report
- e-commerce research report (see the E-Commerce Project in PWOnline)
- feasibility or recommendation report (suggesting design for new web, or redesign of an existing site)
- corporate web site
- oral presentation

> ### Discussion Questions: Corporate Web Project
> The client's particular needs will determine what kind of research you do, as well as what type of project eventually results from your research, as the table on the facing page suggests:

Table 3-4 : Corporate Web Research Questions

Client Needs/Questions	Researchers' Task	Resulting Projects
Should we have a web site?	to determine whether client's products and services lend themselves to the web; to determine cost effectiveness and potential value of site	a) feasibility study that advises client whether or not to implement a site b) new web site design and implementation
Should we upgrade our current site?	to evaluate client's current site (e.g., amount of business generated design evaluation, etc.)	a) evaluation report that gives client a sense of worth/value of site and recommends changes b) web site redesign based on recommendations
Does our current web site meet our clients' and users' needs?	to perform tests on existing site to determine its effectiveness with intended users	a) usability report that gives client a sense of usability of site and recommends changes b) web site redesign based on recommendations
Do companies comparable to our have web sites? How extensive and effective are they?	to collect information about other companies' uses of the web,; to collect sample web pages and compare them	informative report (comparison) analyzing web sites for a particular type of industry

Notes

The primary aims of the Corporate Web Project are to teach you more about

1. how to research and write a major report
2. how to manage a large-scale team writing project
3. how companies and organizations use the web

You should come out of this project with a better sense of how the web works (or doesn't), how large-scale reports should be written, how to negotiate projects with teammates, how to conduct online research, how to survive large and complicated writing projects (most students do survive ... it doesn't have to be horrible and it can even be rewarding).

The Big 1 Case

Reprinted with the permission of Barbara Couture, Washington State University, and Jone Rymer, Wayne State University

Context

As the assistant director for human resources at the Pleasanton assembly plant of the Continental Car Corporation (CCC), you are responsible for helping the plant manager, Frank Page, in a variety of tasks. Today he calls you in to take care of a problem that has him visibly annoyed — probably more annoyed than you have ever seen him.

"The Big-1 Car Rental Agency at the airport has really done it this time," Page says. "Yesterday we had a couple of VPs from a Japanese firm in here, along with an American who represents them out of San Francisco. After lunch, McConkey, the American fellow, and I were walking together out of the restaurant when suddenly he starts telling me about the terrible problems they'd had with the CCC rental car we'd reserved for them at the airport. Seems the heating system didn't work after the first five minutes, so they were freezing during the whole of the 45-minute ride from the Minneapolis-St. Paul airport. But that wasn't all. The car made some funny noises and just wasn't running well. Naturally, the polite Japanese didn't say a word, but Mc Conkey didn't hesitate to say how unhappy and embarrassed he was about the heater not working. Well, if he was embarrassed, can you imagine how embarrassed I was? I didn't know what to say to the Japanese. What a fine way to advertise our products. Yes, sir. Let them see for themselves how wonderful CCC automobiles are!"

You shake your head. "They must have been miserable riding without a heater all the way back from the airport in this 10-degree weather . . . really impressed with CCC's quality!"

Frank Page nods. "And it isn't as if this is the first time that the Big-1 Agency at the Twin Cities airport has given our visitors lousy service.

Several vendors have had problems lately, and have let people in the plant know about it. Why, Manuel Lopez was just complaining to me last week about the poor maintenance on the cars from the airport outlet—some incident with the reps from the Zorelco Company who got a car that had ten things wrong with it. Lopez finally called in and had the agency come out with another car."

"Actually, the complaints have been coming in for several months at least," you reply. "I remember Joe Bomarito telling me some horror story about a high-mileage car from the Big-1 Agency about the time the addition was being put on the front reception area."

"That's a full seven months ago," Page says tersely. "Enough. We have a sufficient history of complaints on the airport franchise. Now we're going to do something about it."

"So what do you want me to do?" you ask. "Go see the manager at the airport agency?"

"No. There's no point in dealing with him. I believe he's proved his incompetence beyond a doubt. Any outfit that consistently rents poorly maintained cars is running a sloppy operation. I want something to happen. The best-made car in the world won't run well if it's not serviced properly. This Big-1 Agency is causing us an image problem."

"And of course we can't take our airport business anywhere else," you note.

"We have no choice," replies Page. "We can only rent CCC products at the Twin Cities airport by dealing with this outfit. That's why I want to blow the whistle on them—make sure that they get their act cleaned up—NOW."

"So we go to the top, to corporate headquarters?"

"Right," Frank Page says firmly. "I want you to write a letter to the general manager of the Rental Division of Big-1 in New York City."

"You don't want to start giving him chapter and verse on our problems, do you?" you wonder.

"No, I sure don't. We haven't kept any records anyway. Look, Continental Car Corporation and the Big-1 Rental Company do a lot of business together. I just want to let the corporate people at Big-1 know that this is an intolerable situation and they'd better do something about it. Get their guys out here to look at this airport franchise See how the place is being run."

"Got it," you reply, as you head back to your desk to draft the letter for Page's signature.

Deliverables

All or some of the following documents might be produced in conjunction with this project:

- a letter to the general manager at Big 1 corporate headquarters, for Frank Page's signature
- a memo to Frank Page
- a letter to the airport franchise manager

Discussion Questions: Big 1 Case

1. What is the problem from Frank's point of view? Are tehre other ways of understanding this problem that Frank might not have considered?

2. What is the best communication strategy to tak in this situation? Should you write the letter that Frank has asked for? Will Frank's strategy solve the roblem -- or should you, in your role as Assistant Director of Human Resources, suggest another strategy?

3. Is it fair to the airport franchise manager not to contat her (or him) before writing corporate headquarters? What are the ethical implications of "going over someone's head?"

4. What reasons might Frank have for asking you to write this letter? How might writing this letter affect your job in the short term? in the long term?

The Kmart Case, or the Renegade Webmaster

Reprinted with the permission of Tim Krause, Purdue University

Context

Your first job after graduation is to serve as Assistant to the Director of Human Resources of Kmart Corporation. After working there for several months, you arrive at work one morning to discover an "urgent" message from the Vice President in charge of marketing. You rush to her office.

"I want you to write a letter of termination for Rod Fournier, our webmaster," Virginia Rago says as you walk into her office, and before you have a chance to sit down. You recall that Rod is Marketing Manager and is also in charge of Kmart's new web site.

"I've already changed Rod's password to block him from the server," she continues, obviously angered more than you've seen her in the past.

You hesitate to even ask what Rod has done before Virginia continues, "Of all the nerve for him to add a link to pornographic material from our web page — Don James called me this morning to tell me about it. I want you to fire him and fire him now. This is bad for Kmart's image. Have a letter with your signature on my desk by lunch time today." She turns back to the stack of work at her desk, clearly dismissing you before you even have time to ask for any additional details.

In the past, Rago has displayed similar outbursts of anger, but you've never seen her quite like this before. You'd go to your immediate supervisor, the Director of Human Resources, but she has the day off. Worse, your files suggest that there is no corporate policy at Kmart regarding the World Wide Web and the information posted there, either by employees individually or by the company. There is a web committee, but you seem to recall that they haven't been active for months.

Your knowledge about the web is sufficient enough for you to know e-mail and Netscape, but you have never viewed the Kmart pages — you've heard they're terribly sketchy and incomplete anyway. You decide that more knowledge about the situation would be helpful before you frame your response.

Deliverables

All or some of the following documents might be produced in conjunction with this project:

- a letter from you to Rod Fournier, firing him.
- Other documents to be determined — what needs to be done? See Discussion Questions below.

Discussion Questions: The Kmart Case

1. What should you do? Should you write the letter that Rago asks for (that is, firing Fournier)? Or should you do something else?

2. Before doing what Rago wants, you should probably ask the following questions:

    - Does the web site create an image problem for Kmart?
    - Is the site problematic enough to justify Rago's response to the webmaster? How do you determine what "problematic enough" means?
    - How have other companies responded to similar cases?

    You expect that tools like Alta Vista and Yahoo might help you answer those questions.

Notes

1 This case is based on an actual occurence. So don't think it is exaggerated or unusual.

2 Further background information about Kmart:

- since this problem occurred, Kmart has created a new page, which you can view at http://www.kmart.com
- you've never been requested to write a termination letter in the past; it's typically the Director of Human Resources job to write them
- there is no union at Kmart

3 Some more general notes about case studies:

- although you want to do some research on-line, remember that the version here is a fictionalized version to give you a role in the case, as are the roles played by individuals in the case in general
- case studies always miss information; keep track of the assumptions that you have to make
- your ultimate goal is to produce a minimum of (1) letter or (1) memo; as appropriate, you may decide to

do more

IF you decide that part of what Kmart needs is a web policy, you can feel free to suggest some direction, but you are not required to write the policy

Employment Project

Context

Your assignment is to locate a job for which you are qualified and apply for it. You can use print or online materials to search for a position — but the World Wide Web is an excellent resource for job postings, at least in some fields. See *Resources for Job Seekers* in PWO.

For most unsolicited applications, you will typically be expected to submit a job application letter (also known as a cover letter) and a print resume. For an entry-level applicant (still in or just out of college), the letter and resume should be one page each. For an advanced applicant or an applicant with extensive relevant experience, the resume might be two pages.

In certain fields (like technical communication, graphic design, architecture), it is also customary to have a portfolio of samples of your writing and design work. You do not mail the portfolio in with the initial application. Rather, if the corporation expresses interest in you, you might bring the portfolio to a job interview.

Increasingly, it is becoming usual to have a web resume: that is, a web page version of your resume (but a web resume can include far greater depth and detail than a print resume).

Deliverables

All or some of the following documents might be produced in conjunction with this project:

- job application letter
- print resume
- web resume

- portfolio
- interview follow-up thank-you letter

> **Discussion Question: Employment Project**
> Who reads your resume and application letter? How do they read these documents?

The Lester Crane Case

by Mark Schaub, March 1998. "Lester Crane: Getting Approvals After the Fact." Business Communication Quarterly. 61-1. 99-111. Reprinted with the permission of the Association for Business Communication.

Context

This case involves numerous decisions on the part of the writer(s): choices of audience, choices related to gender issues, and decisions about cultural differences.

The company

Lester Crane Company is based in Chicago, Illinois and is a privately-owned manufacturer of specialty construction cranes. Exports comprise about 60% of the company's annual gross sales. Some of the kinds of cranes Lester makes include shipbuilding cranes, equipment used in erecting nuclear and conventional power plants, and small mobile cranes used in pouring airport runways. Annual gross sales for fiscal year 1998 were over $980 million.

Your position

You, Susan Russell, are the International Sales Manager in charge of the Middle East market. You work closely with Lester's branch office in the region, located in Cairo, Egypt, and you usually make 5 or 6 trips per year to the Middle East.

Since each country in which Lester does business has different equipment standards and codes, it is part of your responsibility to ensure that the products meet the standards codes of the particular countries. Most ment bureaucrats in charge of determining whether or not a particular piece of equipment will gain government approval — which is vital to Lester's sales.

Specifically, there are two types of government approvals you seek: *Type approvals,* which apply to all cranes of a particular model, and *Machine approvals,* which involve single pieces of equipment, usually specially-ordered, one-of-a-kind cranes.

Not all governments require the same criteria, however, in determining code approval. Some nations, like the U.S., accept actual physical tests of a crane's capabilities and limitations, while others require mathematical calculations of a crane's practical limitations; some governments require both. Still others devise codification's rules on a case-by-case basis, which is why it is imperative that Lester maintain good relations with officials in the bureaus of building and equipment codes in each of the countries to which Lester Crane Co. ships its products.

Your contacts

During your 6 years with Lester Crane Company, you have developed fruitful working relationships with several people, but because of your gender, are still shut out of possibilities in the male dominated Arab countries.

- **Barry Alvarez.** Your assistant, Alvarez handles your office while you're gone. He is a skilled negotiator with a demeanor that appeals to everyone.

- **Latif Abdel-Messih.** An Egyptian employee of Lester, Abdel-Messih is the government liaison officer for Lester's Cairo branch office. He has been instrumental in negotiating the labyrinth of Egyptian national bureaucracy, and seems to know who to talk to about what, often with bribes. You don't ask any questions, because in the past, he's gotten the job done for you — in Egypt, at least.

- **Fritz Canmeyer.** A U.S. government employee, he is the American commercial attache for the Middle East region. Canmeyer splits most of his time between the U.S. Consulate in Jeddah, Saudi Arabia, the U.S. Embassy in Cairo, and the U.S.Embassy in Kuwait City, Kuwait. He has vast experience in working with upper level officials in the Middle East, and has in the past intervened on Lester's behalf to enlist the pressure of government leaders to get your cranes approved. You have been on social visits to Canmeyer's Cairo apartment and you are a good friend of his wife, Judy.

- **Nisham Nazer.** A Saudi prince, he wields great power in his country's government. Though he clearly dislikes dealing with you (you suspect it's because you are a woman), he has been able to win quick approval for Lester cranes in the past. Most recently, he was able to grant acceptance, in less than a week, for cranes needed to complete two new air bases for the Saudi Royal Air Force.

- **Seifallah Abdel-Rahman.** A Palestinian with Jordanian citizenship and the owner of the huge Egyptian-based Arab Contractors, Inc., Abdel-Rahman is a multi-millionaire, who made most of his fortune by using cheap Egyptian or Yemeni labor and inexpensive but low-quality materials to build short-term oil rigs throughout Libya, Egypt, and the oil rich countries of the Persian Gulf. He maintains close connections with Kuwaiti royalty. One of Lester Crane's most significant long-time customers, he has worked with your own superior, Eddie Fogler, for years, and has somehow gotten "official" approval in various countries for cranes that haven't been acceptable for any other government, including the U.S. Eddie Fogler is presently Lester Crane Company's International Sales Manager.

- **Mohammed el-Shafie.** The supervisor of the Jeddah bureau of the Saudi Royal Ministry of Construction and Land Management, el-Shafie seems to be a competent and honest bureaucrat, with some "pull" with the Ministry officials in Riyadh, the Saudi capital. Quiet and reserved during your meetings with him, he seems to always play by the rules.

- **Nasser el-Rifai.** A Saudi citizen, and owner of the sole construction equipment dealership in Jeddah, he has been your largest customer — in the Middle East region — in the past five years. You have dealt with him many times, both in Chicago and in Jeddah. His business has been so successful, largely because his contacts with various Saudi princes has provided him with a near monopoly on bulldozers, earth-movers, and many types of cranes in Saudi Arabia's western region.

Recent history

Lester Crane Co. sold 5 "Sprint 60" mobile mini-booms (worth $110,000 each) to Arab Contractors, Inc., In November of 1991. At that time, your Cairo Office secured *Machine Approvals* for the Sprint 60s, because they were specially ordered by Arab Contractors for use in constructing runway additions at Cairo International Airport, and other major airports throughout Egypt.

The mini-booms are designed specifically for moving and placing prefabricated pieces of reinforced concrete, and were engineered in conjunction with Neuhof AG concrete products of Giessen, Germany, to insure compatibility between the concrete pieces and construction equipment.

After the Persian Gulf War ended in 1991, there was a sharp increase in airbase construction in Saudi Arabia and Kuwait. Late in 1998, Saddam Hussein's renewed buildup of Iraq's military and the increased chemical weapons threat prompted the Saudis and Kuwaitis to accelerate their own plans for increasing the number and size of new airbases.

In December of 1998, the Saudi Government placed a rush order for two dozen of the specially-designed "Sprint 60" mini-booms, placing subsequent orders for concrete prefab pieces with Neuhof, AG. In consultation with Hisham Nazer, the "Sprint 60s" were approved for temporary use in Saudi Arabia, without having to be approved through normal procedures. The cranes were deemed necessary due to the emergency situation. The last cranes on this order were shipped in mid-March, 1999.

In March, 1999, Nasser El-Rifai's company, Gulf Equipment, placed an order for 20 "Sprint 60" model mini-booms. The scheduled delivery date for the first 5 booms is set for June 22, with additional shipments to be made periodically (on July 1 and July 15), until the last 5 are delivered on August 1, 1999. These dates for delivery are specified in the contract, as is the stipulation that securing Saudi governmental approval is the responsibility of Lester Crane. Lester has occasionally re-negotiated contracts and delivery deadlines in the past.

Dilemma

It is June 1999, and 8 of the cranes have been completed at the Chicago plant and are ready for shipment. You, Susan Russell, are in Cairo, Egypt for two weeks, nailing down approval for a different order that will be sent to an Egyptian buyer. Barry Alvarez has taken over most of your Chicago duties while you're away.

The rest of the Sprint 60s are being assembled on the factory floor, when Alvarez receives an urgent email from the Lester expediter's office, telling you that they have no record of codes and standards approval by the Saudi government, and that they need a copy of such an approval before they can begin shipping.

Alvarez realizes that the first cranes were admitted into Saudi Arabia by special governmental decree. Normally, getting such approval means a personal trip to Jeddah or Riyadh, and several weeks of negotiations. He quickly drafts and faxes a letter to Lester's primary contact, Mohammed el-Shafie.

The reply of el-Shafie puts Alvarez in a difficult position. It is your office's job to find solutions to this difficulty. To make matters worse, Alvarez also receives a memo from Jeff Grabill, the Production Manager, that same afternoon. He needs to get an answer as to the status of the Saudi approval as soon as possible. You return from Cairo that night and discover the crisis when you stop by the office on the way in from the airport.

Deliverables

You must determine audience(s) and documents which will help resolve the problem. Once you have determined who you'll write to and for what specific purpose(s), write the documents and accompany them with a Post-Assessment Memo (PAM) defending your primary documents.

You'll probably want to consider drafting at least two documents to two different audiences. You will need to think about writing some memo within the company, and a letter to be sent somewhere outside the company. The audience for at least one of your documents MUST be an Arab or Arabs. You can assume that your audience is fluent in the primary language for international business in Saudi Arabia: English. You are Susan Russell in this case. It would be unwise to delegate your leadership in this instance simply because you are a woman, writing to an Arab audience. A negotiating, high-context culture could easily construe such a move as weakness or (worse) high-handedness.

> ### Discussion Questions: The Lester Crane Case
>
> 1. **Gender.** What aspects of this case, if any, can be attributed to Susan Russell's gender? How can you be sure that gender is treated as inconsequential during the handling of the dilemma? Are there warning signs that you need to watch for? Should Susan Russell be in this position at Lester Crane? Please explain your answers.
>
> 2. **International Argument.** It's common to hear that Middle Eastern argument is more likely to follow a pattern of concentric circles (in contrast with American English which follows a hierarchical pattern that prizes direct, compact statement). Do you see any evidence of other rhetorics at work in these email and letter exchanges? If so, how might that affect your thinking and decisions in the case? If not, do you see any influences of other cultures at work in the documents surrounding this workplace dilemma?

Technology Access Memo/Report for Distance Education Course

by James Porter, Case Western Reserve University

Context

In this project, you are asked to produce a short informal memo/report (about two pages long), written to inform your professional writing instructor about the technology setup you plan to use primarily for your work in her/his course. Such a report is especially helpful in a distance education course, where the instructor needs to know how you plan to access course materials: on what platform? using what web browser? with what e-mail support? etc. Your report should be no more than two pages long (though you might also include an attachment with test results). This memo/report will especially address the often tricky question of file transfer, especially the issue of what happens to files you e-mail to the instructor as attachments. In fact, doing this report will require that you do some *testing* to address questions of compatibility and file transfer (see *Notes* below on testing procedures).

Why does your instructor need this information? Is he/she just plain nosy? No doubt. But there might also be several good pedagogical and strategic reasons for you to write this report. This report is an opportunity for you

- to determine what platform and setup would be most supportive of your work in the course,
- to test file transfer issues, by sending/receiving various types of files to various course locations (for instance, you might test to see whether the class e-mail list will support sending PowerPoint files, or to see how you need to save Excel files in order that your classmates and instructor can open them), and
- for the instructor to know where you will be working, so that he/she can help you solve any file transfer problems that might arise and advise you on whether you have an adequate setup to manage the course from afar.

Deliverables

All or some of the following documents might be produced in conjunction with this project:

- informal memo/report, either print or electronic

Discussion Questions: Technology Access Memo/Report

1. **Where is the best environment for doing online work?**
   The first thing for you to think about: Where is my "home base" going to be? Where will I work, primarily, in this course? Or, will I work in multiple places? In your report, identify the one or two principle locations where you plan to do your work in the distance ed course — places where you think you can work most comfortably and efficiently. Home might be one such place, if you have web and e-mail access there. Or you might identify a campus computer lab as your primary place of work. Why have you chosen this place(s)? *(continued next page)*

Discussion Questions: Technology Access Memo/Report

Don't make this decision in ignorance. Don't just speculate. Rather, do some *testing* to determine where you should work and how you might need to adjust your work space or habits to meet the demands of the course and to coordinate with your instructor's and classmates' technology setups. *Your written report should provide test results!*

Questions to reflect on and answer:

- What machine will you be working on? using what operating system?
- What word processing program will you use?
- Are you able to access the World Wide Web from your home computer? Can you load web pages quickly and easily? Does the web-based chat server run efficiently? (Hint: you might run a timed test to see how long it takes to load the chat page on your home computer. You might also test out with some other students how efficiently chat will work for you at home.)

2. **Will this environment support file transfer in this course?** One important issue for a distance education course concerns file transfer within e-mail messages. Run a few "file transfer tests" to determine what problems you migh face with file transfer between you, the instructor, and other members of the class. Send some test message to determine answers to questions such as:

- At this location, using this machine, can you receive attached files from the classlist? from the instructor?
- Willyour attached files be easily downloadable by the instructor? by other students? in what format do you need to sve files in order to send them to others?
- Have you tested different kinds of files? (e.g., Word files, PowerPoint files, Excel files)

- appendix of test results (in table form)

Notes

Suggested test procedure for file transfer

The main focus for this early assignment should probably be word-processed files sent as e-mail attachments. Here is a suggested testing procedure.

1. Create a small sample document in your word-processing application. Include in that document a variety of features, including a table and some graphic elements (like various font sizes and styles, tabs, bold type, italics, horizontal rules).

2. Send that file to your professional writing instructor (and/or to the class list) as an attachment to an e-mail message. Ask the instructor and other participants in the course to report back any problems that they have with accessing that file. (Keep a record of your results see Table 1.)

3. If there are problems in step #2, then try sending the file again — this time saving it in a different format. The goal is for you to find out what file format works best for mailing projects to your instructor and to the class.

See "Notes" in the main Technology Access Memo/Report page on the website for more details on testing procedures, tips for organizing and formating the memo/report, and setting up tables.

NCAA Baseball Bat Standards: Complaints and Policy

by Patricia Sullivan, Purdue University

Context

You are a new intern in the sports medicine research department of the NCAA (National Collegiate Athletic Association). Although you are interested in swimming and diving, you are assigned to answer complaint letters as a first task. When you complain that you know nothing about baseball and could care less (though you state that very diplomatically), your supervisor Mike Smith says that everyone is taking a share of these letters because there are more than 100 letters to answer and the office staff is not equipped to explain the science. You ask whether some templates have been developed for the types of complaint letters so that similar scientific responses can be used and the bulk of the complaint letters can be handled more efficiently. "Good idea," Mike responds, "See if you can develop some templates. That will help us a lot."

Great. You don't know or care that much about baseball and you are the person who has to develop templates for responses to queries on this issue. Not only that, Smith makes it clear to you that he is waiting for your templates before he answers his share of the letters, and he will judge your early work on how well you solve this department problem. "If you can just get us some good templates," Smith tells you, "we can get back to the work of the department." Some people do know more than you do, so you quickly realize that you will have to gain agreement of those others in the department if your work on templates is to actually to be used. Your tasks in this situation, then, are to classify the letters, get others to accept those classifications, and to build suitable templates.

Research

To find out more about the background you talk to some of the staff interested in baseball and bats. From them you assemble this story:

Since Mike Mussina (pitcher for the Baltimore Orioles) was struck above the right eye by a batted ball in Spring of 1998, the public has viewed bat technology unsympathetically. Perhaps because Mussina's injury came from a balled batted from a wooden bat (which does not generate the speed that composite and aluminum bats generate), perhaps because the incident was incredibly scarey and was replayed on ESPN for weeks, the NCAA made a move to join the SGMA (Sporting Goods Manufacturing Association) in the testing of bats.

The key questions for bat testing research were:

- how fast is a batted ball moving when it reaches the pitcher? *and*
- how quickly can a pitcher react to a batted ball?

Deliverables

You need to complete two documents for this project:

1. **Preliminary Analysis Memo.** This memo states the situation, gives your approach, and provides the results of your preliminary research into the plausible responses to these letters. Though it is addressed to your supervisor, Mike Smith, it will be distributed to the department and be the basis of a

departmental discussion about the best explanations to use in the responses.

The research should be presented in table form as a way to facilitate department discussion.

2. **Template Plan.** The second memo will be used by the department to respond to the letters. It should identify the letters by type of complaint and include a revised table that: (1) identifies one or two reasonable responses to each type of letter, (2) includes wording for key explanations, and (3) provides a source for information that backs that explanation.

The template plan should also include a completed letter that people can copy.

Department meeting

After you have completed your Analysis Memo the class (or some other group of people) will meet as a department. At that meeting the group will examine the possible responses that have been generated in the Analysis Memos, and they will reach some compromises about how the questions should be answered.

That Department meeting should generate a guide for developing the templates.

Discussion Questions: NCAA Bat Standards

1. Roles of the writing plays: The planning worksheet supports the suggestion that you can view this project as producing text that aims to achieve different goals: you might see that text as teaching the people who wrote to the NCAA, or as defending the position of the NCAA to the people who wrote, or as selling those who wrote on the interpretation that fits with the NCAA decisions about bat standards, or as informing those who wrote of the decision on the NCAA. What roles do you think are more and less important? [give a couple of reasons why you think as you do] Are any of the typical writing roles irrelevant to this project? [if yes, why; if no, why not?] *(continued next page)*

2. **Your role in the organization:** Have you held a job as an intern or as the new person in a working group and had to speak with authority when you really felt you had little? That certainly is the position this intern finds herself/himself in at the NCAA Sports Medicine Research Department. What strategies can the new person who is forced to speak with authority use to build respect and acceptance in the group? Which have worked better (and worse) for you?

3. **Letters serving as data:** Examine the letters you need to use as the anchors for your templates. Which of them are easier to answer and which of them are more difficult to answer?

Software Bug Case

by Colleen Reilly, Indiana University - Kokomo

Context

You are a product manager in the marketing department of a large manufacturer that builds electrical meters and metering products for sale in the United States and abroad. One of your largest customers recently called you to report that they found a bug in the newest version of your company's metering software. This software, called MS2000, is used to program meters in the field. According to the customer, the bug causes a meter to reset itself to zero at a certain point, resulting in an inaccurate calculation of electrical usage.

You have performed tests on the latest version of the MS2000 software, version 3.13, and discovered that the bug surfaces when the software is used under the conditions described by your customer. You realize that this bug might also pose a problem for other customers.

In conjunction with your colleagues and your group leader, you decide that you need to draft a letter that can be sent out to all customers who might be using the newest version of the MS2000 software. You are assigned to write the letter because the software product is in your area of concern.

In writing the letter, you need to consider the following facts and customer relations concerns:

Facts about the situation
- version 3.13 of the MS2000 software will fail under the conditions reported by the customer
- version 3.13 of the MS2000 software may fail under other conditions not yet discovered
- software liability laws require that you inform customers about known software problems and make it easy for them to obtain fixes and technical support
- best option for customers is to return to using a previous version of the software and then to update to another, Y2K compliant version. In order to do this, the customer should copy their backup database into the ms\ms2000\db directory to restore the programs they were using previously and obtain the Y2K compliant version of the MS2000 software, version 3.03C Y2K, and perform an update load of that version

Customer relations concerns
- you do not want to cause customers to lose confidence in your products and/or begin looking for other problems with the software that are outside of their current uses
- you need to be careful not to cause customers to panic unnecessarily; however, you want all of them to be sufficiently motivated to stop using the 3.13 version of the software
- you should avoid using imflamatory language, such as the word "bug," when speaking to customers about the situation

Deliverables

Planning/analysis of the case

Your first assignment is to analyze the major issues in the case and outline the major concerns of the metering company and the customers. You should draw on the background information in the case and your research concering software liability. You should make lists of possible solutions and approaches. Any solutions must involve assuming a position with regard to software liability.

For this part of the assignment, you will submit a memo to your group leader containing a statement of the main issues in the case, including information about software liability.

Letter to customers

You will compose the required letter to customers. You should use correct business letter format.

This letter should reflect the careful planning and analysis of the case done in the first part of the assignment. The letter should be general enough for all customers but should be strong enough to prompt affected customers to act. Your letter should clearly explain the fix that customers should use to protect themselves from problems with the software. The steps that customers must take are outlined above in the facts about the situation.

Discussion Questions: Software Bug Case

1. What is the context and purpose for the analysis memo?

2. Who are the possible audiences for that memo?

3. Based on your research, what are some of the software liability issues that need to be considered when writing the letter to customers?

4. What is the best way to explain to customers what they need to do to address the problem?

5. What should you avoid mentioning in the letter mention that could harm or compromise the position of your company? How can you both protect the company and behave ethically with respect to your customers?

The Job Banks Project
Patricia Sullivan, Purdue University, and Teena Carnegie, Oregon State University

Context

The student chapter of your future professional organization has heard positive and negative comments about electronic resume and job searching services (or job banks). When you ask your chapter advisor, Professor August Anselm, about which online employment services to use, he turns the question back on you. "What a great chapter project to get the semester started and everyone contributing," Professor Anselm responds. "You can create some guidelines for what it would mean for an online sites service to cover our discipline particularly well, and then rate the various web sites. You can even develop some materials for students and a workshop presentation. Of course, you will have to show everything to a couple of the faculty, so that we can be sure that your research and conclusions about which job banks to use reasonable ones. Once you're ready to present it to the chapter, I'm willing to bet you that it will be the best attended meeting of the year. . . We might even add your work to the department website in the student services section."

You are not altogether pleased about this assignment because you plan to go to graduate school rather than pound the pavement. Also, you suspect that showing the materials to a couple of the faculty may require you to revise much of your original work. But, Professor Anselm has held out rewards that attract you. You know that running a professional workshop is excellent experience to gain as a student (it recommends you as a management trainee and as a potential teacher). You also would like to have materials on the department web site with your name attached to them; such a credential might increase your grad school opportunities. Also, several of your friends are excited about the project, so it may turn into a fun resume item.

Realizing that you know little about these services, that you plan to go to grad school rather than look for employment next term, and that it is hard to predict the time involved, you get the group to agree to do some investigation before committing to the project.

To begin your investigation, you visit with the department counselor, Regina Witherspoon. She offers a number of important pieces of information: (1) that all job banks are not the same — some work with experienced workers, some specialize in particular fields, some allow you to post a resume, most require you to fill in a form that they electronically post; (2) because most control the categories for information, you often need to rethink your resume in order to fit it into their formats which sometimes favor other types of experience; and (3) that not all are as successful in placing candidates.

She then shows you job-hunt.org, a clearinghouse of job searching information, as a way to examine the categories. "I don't want all these sources to intimidate you," she says. "Let's look at the ways that they organize the types of information. "Clearly there are hundreds of sites listed here. Note the language. We have been talking about 'Job Banks' and at least this site lists those sites as 'Employment Super Sites' and as 'Resume Banks.' As you search for information on the web, you may need to pay attention to finding the right words. "I also caution you to note the key criteria —i.e., who pays? Most sites allow applicants to post free and charge employers a subscription fee for accessing the posted resumes. You want to stay away from third party sites where the job seeker pays 'Recruiting Agencies' and 'Commercial Services' (and we might add 'Headhunters'). Students have already paid to use the university's career placement, so paying commercial groups becomes a later option."

Ms. Witherspoon also suggests you begin by making a list of criteria for what would make it a good one for your major, reading the online articles about job banks, and looking at the candidates for inclusion.

Deliverables

All or some of the following documents might be produced in conjunction with this project:

Phase 1
Preliminary analysis. If you follow Professor Anselm's and Ms. Witherspoon's advice, you need to:
 1. develop some criteria
 2. search the web and develop a list of available sites
 3. find information about how to judge the sites — the faculty will

accept published opinions
4 look around at the candidate sites;
5 write a preliminary analysis memo that presents your recommendation about whether to proceed with a workshop or instructions or web materials about appropriate job banks for your major.

By way of backing for your opinions, the analysis memo should include links to research, to good banks, and to bad job banks. Should you decide a workshop on job banks is not the best move for your major, you should also decide whether there are some useful web resources to publicize (e.g., job listings in the area, newspaper classified pointers, lists of job sources, and so on).

Oral presentation (optional). Your teacher may want you to present your findings to the rest of the class orally as a way to build the class' understanding of Job Banks. In that case you will need to work with your preliminary analysis to develop

1. a handout on Job Banks for your major (this may be a table of analysis), or
2. a computer-based presentation that shows some aspects of Job Banks

---

Discussion Questions: Job Banks Project

1. Your decisions about the usefulness of job banks will have input from your opinions of the job banks and your opinions about the usefulness of online searching in your major. Before you start, share any information you already have about online job searches with your classmates. Have you (or a friend or family member) used the web for job searching? If so, what happened, and how does that shape your opinions for this project?

2. If an online resume service required it, would you put your picture into your online resume? Why (or why not)? Would you wonder how the picture was being used by prospective employers.

Notes

If you continue this project past the deliverables produced during the initial assessment of job banks for your major, click to

- Phase 2: Job Bank Interface Comparisons
- Phase 3: Job Bank Workshop

National Electric: Creating Visuals for Employee Statistics

by James Porter, Case Western Reserve University

Context

This one-day assignment asks you to design a table or visual and justify its use in communicating employment data to your readers.

You work for the public relations department of a major electric company, National Electric. Recently the utility has been criticized in regional newspapers — both in editorials and in letters to the editor — for pushing a substantial percentage of its older workforce into early retirement. The result of this bad press has been the perception that the company lacks experience and maturity and that its workforce is now young and inexperienced.

The director of public relations, Susan Smith, plans to write a short statement rebutting this view and to publish it in the newsletter that gets mailed to customers with each month's electric bill. This statement will include a visual illustration or table providing data about the years of experience of the electric company's workforce. The visual will be published in the company newsletter — but it will also be attached to a press release sent to local media (newspapers, television news) and included on National Electric's corporate web site.

Data

As of January 1, National Electric had a total of 7,863 employees. Of these, 1,989 (or 25%) had less than five years' service; 1,275 (or 16%) had ten to 15 years' service; 784 (or 10%) had fifteen to twenty years' service; 931 (or 12%) had twenty to thirty years' service; 1,294 (or 16%)

had thirty or more years' service with the company.

Deliverables

All or some of the following documents might be produced in conjunction with this project:

- a memo to Susan Smith, the Director of Publications for National Electric
- a table or visual appropriate for the task

Discussion Questions: National Electric

1. Is the employment data appropriate? Is the data acurrate? (Did you check the calculations?) Are there other forms of data which would make the point more effectively? more honestly? Is more research required?

2. What kind of table or visual would best convey the information National Electric wants to pass on to customers? table, pie chart, bar graph, line graph -- and with what heading or textual features?

Notes

1. *You have numerous options in presenting the data. You can report numbers only, percentages only, or both. You do not have to present the numbers as they are listed above; you could combine categories.*

2. *One strategy in writing the memo would be for you to (a) briefly summarize the problem facing National Electric, (b) indicate what purpose the visual or table ought to serve in informing and persuading the public, and (c) justify your choice of graphic.*

3. *Your goal is to present the company in the best possible light — but you also have to represent the data accurately and fairly so as not to manipulate or deceive your customers and the public.*

4. *The visual or table should be simple and easy to read, since it is going out to the general public. It should be clearly and concisely labeled. It should convey the message directly and with some persuasive impact. Of course it should be attractive and professionally designed.*

E-Commerce Project

by James Porter, Case Western Reserve University

Context

In this project, you are asked to do web-based research (and perhaps some library research) and to write an informative report on the question of electronic commerce, or "e-commerce." Is it a good idea for businesses to invest in electronic commerce — i.e., use the Internet and World Wide Web as a medium for promoting their corporate image or for selling products and services?

Before you shout "YES, it's the web, stupid!," think about the question in a serious way. First, this is a broad question, and even if you think the answer should be "yes" (or "no"), your first step is to focus thoughtfully on the issues. What are the advantages and disadvantages of e-commerce? Why would somebody want to have a business presence on the web? What are the problems involved? What about costs — and are the costs of a web site worth it in terms of return on investment?

Your research will result in a short (2-3 pp.) informative report responding to the question you raise. This report will be a "generic report" (like a white paper) that is not intended for a particular company or organization but that focuses rather on the broader issue of e-commerce. The purpose of this project is (a) to help you gain some expertise on a particular topic of importance to business (web commerce), and (b) to produce a report of interest and value to businesses that might be considering engaging in web commerce.

Who is the audience for this project? You could imagine you are part of a consulting firm, like Jupiter Communications, that prepares reports for various businesses, and write this report as an a kind of magazine article that could be read by any number of business people interested in the question you raise.

Deliverables

The following document might be produced in conjunction with this project:

- informative research report (print or PWOnline)

Discussion Questions: E-commerce Project

Research questions. Your first task will be to do research to educate yourself on some of the general issues regarding e-commerce. As you read, begin to focus on a particular issue or question of interest to you that you also think is significant for businesses to consider. You might, for instance, decide to focus on business use of the web in the florist industry. Do florists tend to use the web? Does it work well as a vehicle for e-commerce? what are the particular problems they face selling their products online? Here are some specific and legitimate questions that businesses might have:

1. Are online credit card transactions secure? Do customers feel safe using credit cards online?
2. Is web advertising worth it?
3. How much does it cost to sell products via a World Wide Web site? How does a business go about creating such a site?
4. Which products and services sell well online? Which don't?
5. What types of businesses and organizations are best served by e-commerce?

Notes

Organization, format, and design of the report

- **Header.** Provide basic information about your report: title, author name, and date.

    Are Online Transactions Secure?
    A Report on E-Commerce and Credit Card Use

    Roger Thorndyke
    March 19, 1999

- **Overview and opening paragraph.** Indicate what your report is about — what question or topic does this report cover? what is the scope of coverage? And what is your overall thesis or main point?

- This report examines the question of whether credit card transactions on the World Wide Web are actually secure. Though there have been many instances of online credit card frauds, the incidence of such frauds is actually no more than the regular plastic transactions. However, there are important steps a company should take to insure that its customers' credit card transactions are handled securely.

- **Headed subsections.** Divide your report into subsections, depending on the major topics you will cover in the report. If your report addresses the question of whether a bookstore should move to online, you might divide your report into two possible major sections: Advantages of Web Book Sales, Disadvantages of Web Book Sales. You might also have sub-subsections which distinguish between various advantages and disadvantages.
    - Frequency of Online Credit Card Fraud
    - Steps Companies Should Take to Prevent Credit Card Fraud

- **Graphs or charts.** Think about how to have visual impact in your report. A line graph could be used to show change over time — for instance, the increasing number of businesses engaged in web commerce, or the # of incidents of online credit card fraud. If you borrow a chart or graph already contained in another source, be sure to identify that source below the chart.

- **References.** Provide a list of the references you use in the report, as a separate final page of the report. Be sure to format your individual entries according to APA format (or whatever format is specified).

Note to teachers

In many respects, the E-Commerce Project is a smaller version of the Corporate Web Project. However, there are some important differences: the E-Commerce Project is intended to be done by individuals rather than teams; it generates a more traditional research report (in many respects similar to an academic term paper); and it is focused on a broader question of interest to multiple companies (whereas the Corporate Web Project is usually intended for one client company). The E-Commerce Project is also intended to teach students how to do web-based research: how to read online materials critically; how to synthesize information from multiple sources; and how to write a well-organized and useful report based on, mainly, an analysis of published material. In essence, the project teaches the skills of writing a business research report. This project is recommended as a good warm-up exercise for the Corporate Web Project —

chapter 4

Documents

The *Documents* section of the Professional Writing Online site provides templates and samples for various types of professional documents and genres. It also provides links to the *Principles* sections where these documents are discussed and explained, and links to some of the *Projects* which use them. The *Documents* section addresses the following types of documents:

- Memos and E-Mail
- Letters
- Reports
- Policy, Manuals, Handbooks
- Employment Documents
- Promotional Materials
- Instructional Documents
- Oral Presentations

This section of the Professional Writing Online Handbook contains examples of student- and professionally-written documents included in PWOnline.

The sample documents you'll find in this chapter include:

Sample letters	Sample memos
• Request for Deposit Refund • Revised Request for Deposit Refund	• Announcement of Policy Change • Revised Announcement of Policy Change • Project Planning Memo • Revised Project Planning Memo
Sample reports	Sample employment documents
• Corporate Web Project Proposal • Corporate Web Recommendation Report • Documentation Project Plan	• Application Letter: Research Position • Application Letter: Teaching Position • Resume: Teaching Position • Reference Sheet

Sample Letter - Request for Deposit Refund

By Christina Rinderle on behalf of four tenants. Reprinted with the permission of Christina Rinderle.

606 Evergreen St.
West Lafayette, IN 47906
765-746-1000

10 June 1997

Mr. Greg Beason
Purdue Research Foundation
1220 Potter Drive
West Lafayette, IN 47906

Dear Mr. Beason:

On June 3, 1997, we were happy to promptly receive our security deposit from 240 West Wood Street; however, we were upset to see each of our checks $32 shy of the full amount. This totals a $128 loss for a house that was left in not only acceptable condition, but far better than that when we moved in.

We realize the situation at this house is an unusual one, but we, the tenants, should not be punished for circumstances beyond our control. Last year we each signed a lease with Mr. Brian Lane of Lafayette. We were shocked to find out that our contract was broken in the spring of this year when Lane sold the house to PRF.

When we contacted PRF on the afternoon of June 3rd, we were informed that our loss was due to a "dirty refrigerator, filthy oven, disgusting bathroom," and that the walls needed to be repainted. We had disinfected the refrigerator, cleaned the oven, and scrubbed the bathroom. As for the walls, they have not been repainted for at least 4 years, and dirt marks that you may have noted are not above normal wear. Is it not standard practice that you provide new tenants with freshly painted walls anyway?

We are asking that you maintain your prompt standards in replying to us and enclose checks for the remaining $32, payable to the undersigned names. Thank you.

Sincerely,

Christina Rinderle
Kathleen Biddle
Andrew Larson
Gretchen Thompson

Sample Letter - Revised Request for Deposit Refund

By Christina Rinderle on behalf of four tenants. Reprinted with the permission of Christina Rinderle.

606 Evergreen St.
West Lafayette, IN 47906
765-746-1000

10 June 1997

Mr. Greg Beason
Purdue Research Foundation
1220 Potter Drive
West Lafayette, IN 47906

Dear Mr. Beason:

On June 3, 1997, we were happy to promptly receive our security deposit from 240 West Wood Street; however, we were upset to see each of our checks $32 shy of the full amount. This totals a $128 loss for a house that was left in not only acceptable condition, but far better than when we moved in. This loss may not be a significant one to PRF, but it is to us as college students.

We realize the situation at this house is an unusual one, but we, the tenants, should not be punished for circumstances beyond our control. Last year we each signed a lease with Mr. Brian Lane of Lafayette, which stated that our deposit would be refunded if we left the house in acceptable condition. We expected PRF would continue to honor these terms upon acquiring ownership of the house.

When we contacted PRF on the afternoon of June 3rd, we were informed that our loss was due to a "dirty refrigerator, filthy oven, disgusting bathroom," and that the walls needed to be repainted. We had disinfected the refrigerator, cleaned the oven, and scrubbed the bathroom. As for the walls, they have not been repainted for at least 4 years, and dirt marks that you may have noted are not above normal wear. In addition, according to the lease we signed we were not responsible for repainting upon moving out.

In light of this, we feel we have fulfilled our obligations and have not violated the terms of the lease we signed. Thus, we are asking that you maintain your prompt standards in replying to us and enclose checks for the remaining $32, payable to the undersigned names. Thank you.

Sincerely,

Christina Rinderle
Kathleen Biddle
Andrew Larson
Gretchen Thompson

Sample Memo - Announcement of Policy Change

Purdue University

TO: Graduate Instructors in Professional Writing
FROM: Pat Sullivan
DATE: July 7, 2000
SUBJECT: Evaluations of English 420 and 421 Courses
CC: Dean Rowe, Dean Gentry

Because of remarks that aim to harass and intimidate instructors, we are changing procedures for course evaluations. This goes into effect immediately.

You should still use the special CIS form for the Professional Writing Program, but now you will also instruct students that they must sign their names if they write comments on the back. Yes, there will be no more anonymous written comments returned to teachers.

People have said that such a move limits student freedom, but it seems to me that freedom is always tempered when it threatens others in the society.

We will discuss this further in staff meeting. Thanks.

Comments
This draft rambles and is emotional. More specifics are needed to give the teachers some idea of the reason for this move. The memo also is too informal in tone to send to the Deans that are cc-d.

Sample Memo - Revised Announcement of Policy Change

Purdue University

TO: Graduate Instructors in Professional Writing
FROM: Pat Sullivan
DATE: July 7, 2000
SUBJECT: Evaluations of English 420 and 421 Courses
CC: Dean Rowe, Dean Gentry

Effective immediately English 420 and 421 will change their procedures for collecting course evaluations.

Old Procedure

1. order CIS for the Professional Writing Program
2. read procedures to students which encourage them to write comments on the back of the form and which assures them that the comments will remain anonymous
3. leave room and have a student collect the forms and return them to the department office

New Procedure

1. no change
2. read procedures to students which encourage students to write comments on the back of the form. Stress that threats of violence are against the Purdue code of conduct and can result I disciplinary action. Thus, all written comments must be signed (including ID#)
3. no change

This change is deemed necessary because of one student's written comments that aimed to harass and intimidate her/

his instructor. An anonymous written comment indirectly threatened one of our instructors this Spring, and such threats will not be written to our staff again. Although this action limits student freedom, freedom is apt to be curbed when it threatens others in the society.

We will discuss this policy at staff orientation in August. At that time we will also discuss ways to help students ponder what constitutes harassment in a workplace environment.

Sample Memo - Project Planning Memo

Purdue University

TO: Ms. Mary Lisa Romney
FROM: Sharon Case, Arnette Jones, and Jae Kim
DATE: June 15, 1999
SUBJECT: Web Companion for New Students

As part of the Lilly Grant to Improve Retention of Students the Summer Orientation Program wants to build a Web Companion Site for New Students 1) that connects them to local information they need, that helps them get oriented online, and 2) that provides online help for typical web tasks and software. For our web solutions project we plan to develop a site plan for such a web companion (including some kind of cute Purdue name) and a sample piece of documentation.

We think that students will respond better to a web companion that reflects their view of themselves, so we are going to study whether we can make a number of personalized splash pages that are possible (by setting a profile).

We are going to research the following groups:

- International Students
- Engineering Students
- Agriculture Students
- Liberal Arts Students
- Business Students
- Technology Students

The research will be used in a report about how to organize

the site.

Then we will select one of the groups and develop a splash page idea and site plan that tailors the site to their needs. That splash page and site plan will be submitted to other teams for review. After we have our approach critiqued and approved, we will develop the splash page, the site map, and some sample copy.

We will present the results to the class and have them try out the new site.

Comments

This memo has appropriate content for a project planning memo because it tells about the occasion, the project goal, the insight that drives the work, who will be studied, and the steps in the process. It does not identify the writers' relationship to Ms. Romney, nor does it explain why (or how) they are researching the groups they are researching, nor does it give a schedule for the work.

Sample Memo - Revised Project Planning Memo

Purdue University

TO: Ms. Mary Lisa Romney, Project Director, Lilly Retention Grant
FROM: Sharon Case, Arnette Jones, and Jae Kim, English 421 students
DATE: May 25, 1999
SUBJECT: Project Plan for Researching the Feasibility of Developing a Web Companion for New Students

As part of the Lilly Grant to Improve Retention of Students the Summer Orientation Program wants to build a Web Companion Site for New Students 1) that connects them to local information they need, 2) that helps them get oriented to online information, and 3) that provides online help for typical web tasks and software. For our course project in English 421 we plan to develop a site plan for such a web companion (including some kind of catchy name) and a sample splash page for that site. Our project will identify the sorts of content that students want to see on a Purdue Web Companion and investigate whether we should tailor the site to specific groups of students.

Solution We Will Investigate

When we conducted some preliminary research of web companions at other universities, we found that a number of them tailor the sites to specific student groups. Thus, we think that new Purdue students will respond better to a web companion that reflects their views of themselves. Because this idea would require some special programming, we recognize that it needs to be researched. So we are going to investigate whether we can make a number of slightly personalized splash pages that can possibly (by setting a profile) make new students feel more welcome at Purdue. We will interview first or second year students from the following groups:

- International Students
- Engineering Students
- Agriculture Students
- Business Students
- Technology Students

Their responses will confirm our solution. In those interviews students will be asked to respond to a variety of companion sites, telling what features they like and dislike. They will be shown some sites that are aimed at particular groups and some sites that are more general. The goal of this research is to have them 1) respond to those sites and 2) identify what features would help new students at Purdue.

Deliverables

The interviews will be the basis of a report about how to organize the site.

The rest of the project will plan the site that the interviews support. We will produce:

- a site plan — an outline of the contents of the site
- a splash page — a sample opening page for the site
- a content page — a sample of some content that students want included

The splash page and site plan will be submitted to other teams for review. After we have our approach critiqued and approved, we will develop the splash page and some sample copy.

We will present the results to the class, have them try out the new site, and make final adjustments.

Schedule

We propose the following schedule for this project:

- May 25 Project Plan submitted
- May 28 Interview questions reviewed
- June 4 Interviews completed and preliminary report of research drafted
- June 8 Splash page ideas and site plan ideas reviewed
- June 11 Site plan and sample content drafted
- June 15 Oral presentation of findings and materials
- June 18 Project memo plus revised Site plan, sample content and splash page

We welcome your comments and look forward to completing this important project.

Sample Report - Corporate Web Project Proposal

Web Page Solutions — Memorandum
To: Leonard Nimoy, Manager
From: James T. Kirk, Research Analyst
Date: May 22, 1998
Re: Proposal— Web Project for The Imagination Station

Prospective Client
I believe that The Imagination Station would make a great client. The Imagination Station is a "hands on" museum for children that is open on Friday, Saturday, and Sunday. This client is located here in Lafayette. What drew me to this particular client is that it is a non-profit organization. The Imagination Station survives on donations and contributions. It is a small organization with minimal funding. Many of the workers at the station are volunteers. One of the founders is a woman that I work with at the Visitors Information Center at Purdue. She currently is President of The Imagination Station. Her ease of access makes the Imagination Station a strong candidate for the web project.

The Client Need
The Imagination Station is currently looking into the benefits of a web page on their own. The cost of undertaking and maintaining a web page has been their main concern. As mentioned before, the Imagination Station has a fixed budget. The benefits that The Imagination Station can foresee are:

- increased visibility within the area
- a systematic way of interpreting their target audience
- simply keeping up with the times
- a timely way of receiving feedback

The client desires to increase its visibility to Lafayette and other neighboring cities. The client realizes that most organizations are online these days, and they would like to stay up to date as well. The client also realizes that more and more children have access to the Internet. By being online, they will be more visible to its target audience. They wish to provide an online questionnaire which would allow them to receive feedback as well as e-mail addresses.

Qualifications

This project can be done within a reasonable amount of time. All of the resources that are needed can be found here in Lafayette. Annette Goben is the woman that I work with which I referred to earlier. She would be more than happy to work with myself and the team on this project. As I mentioned earlier, The Imagination Station has already looked into a web page. This project could both benefit The Imagination Station as well as being a learning experience for the class. What I can bring to this project is minimal. I do not have much knowledge in the area of the web, and can not help out in this department. I can conduct the research and would be more than happy to be the liaison. I hope that this project will provide me with web experience.

Research Resources

The client would be our first resource. They can provide us with their ideas of what they would like on their web page. Due to the fact that they have already looked into developing a site, they have an idea of what they desire right now. The Imagination Station would be more than happy to help us with this project. Visitors to The Imagination Station would also be a source of information. They can provide us with details that would be helpful for them to see on the web, such as directions, things to see, and cost. Another resource is a friend of mine whose job is to develop web sites for organizations. I do not know at this time of any other organization quite like The Imagination Station which could provide us with additional information.

Recommendation

I believe that this organization should be considered for the web project. It is feasible to undertake due to the fact that the client is so close. The organization could benefit from this project in many ways. One benefit, is that The Imagination Station has a need for a web page. The second benefit is that the research is free. Another benefit for them is that this type of research can be fairly time consuming. The Imagination Station is also an interesting client. They developed an exciting learning environment for children with limited resources. My concern is whether or not this proposal will support three or more people. I strongly encourage Web Page Solutions, Inc. to consider The Imagination Station for this project.

Sample Report - Corporate Web Recommendation Report
Reprinted with the permission of Christina Rinderle, Dave Ong, and Brent Willman.

Memorandum

To: Dr. Bertin Anderson, President, RWMC
From: Christina Rinderle, Brent Willman, Dave Ong,
Research Analysts, Web Page Solutions
Date: June 12, 1997
Re: Recommendation Report: Design and Benefits of a Web Site for RWMC
Cc: Jim Porter, Manager, Web Page Solutions

Overview

The purpose of this recommendation report is to propose the design of a web site and discuss the benefits of having a web site for Revegetation and Wildlife Management Center (RWMC). Web site technology has become increasingly popular with many small businesses in recent years, especially among many of RWMC's competitors. Our team has been approached to investigate two issues: (1) design of a web site for RWMC to specifically highlight RWMC's services, pricing policy, revegetation techniques, and previous projects undertaken, and (2) the benefits of having a web site for RWMC in terms of publicity and competitiveness.

We will first present an executive summary of the results of our research for the above mentioned two issues, plus our recommendations. This will be followed by a description of RWMC and its needs. We will then discuss our research methodology and examine in greater detail our findings. The findings will be analyzed and presented in our recommendations. A list of the appendixes can be found at the end of the report.

Executive Summary

Findings: Design, Cost, and Benefits of a Web Site RWMC is keen on the prospect of having a web site. This report examines the following findings concerning RWMC's need to have a web site:

- Design of RWMC's web site should at least include:

    - credentials of staff
    - revegetation techniques
    - previous projects undertaken
    - pricing policy for services and materials

- Costs to setup and maintain a web site

    - approximately $1,029.35 to setup the web site for Year 1
    - approximately $409.40 to maintain the web site for Year 2

- The extent of web site to promote publicity and competitiveness is limited unless web links are:

    - established under a set of keywords with various search engines, eg. Yahoo and Infoseek
    - established with previous clients of RWMC
    - established with suppliers of materials and native seeds to RWMC
    - printed on brochures of RWMC to be handed out

- Design of RWMC's competitors' web sites:
  - More similar features include:
    - reclamation techniques
    - reclamation goals/results
    - capabilities of firms
  - Less similar features include:
    - credentials of staff
    - photos of sites (Before and after
    - revegetation)
    - prices of services, materials and
    - native seeds

Recommendations: We conclude with these recommendations:

- We favor the prospect of a web site and itshould be hosted by a professional web design company. We recommend BestWeb 2000, in terms of their best quality and quantity value for money.

- RWMC should at least include in the design of its web page the credentials of its staff, previous projects undertaken, revegetation techniques, and pricing policy.

Description of RWMC

RWMC. RWMC, located along the Lower Colorado River, focuses on restoration of riparian habitat types. Such habitat has been destroyed mainly by continued overuse of the Colorado River and/or increased urbanization. Primarily, RWMC1s clients are complying with mitigation requirements or seeking to create preserves for species that are threatened as a result of such destruction. Efforts to restore these areas include soil analysis and irrigation installation, followed by replacement of native plant species and post-restoration monitoring.

RWMC's Needs. RWMC needs to promote its publicity within Colorado and the various other states. With 20 years of experience in revegetation, RWMC has developed a unique 9-step revegetation plan and performed services for numerous clients. However, publicity has mainly been by word of mouth and local knowledge. With strong competition from rivals with web pages, we strongly believe that a web page would be ideal for RWMC to more efficiently and effectively provide their services, and put RWMC on equal grounds to compete with its rivals.

[Full report not yet included on site...See sample report on web site for complete appendix list and bibliography for this sample report.]

Sample Report - Documentation Project Plan

Reprinted with the permission of Adam Banks, Andrea Cumbo, Margaret Fulton-Mueller, Naomi Iganashi, Amy McAlpine

The following project plan shows how a team of five students organized its work for the Documentation Project over an eight-week period. For a copy of the Gantt chart for this project plan, see the sample in PW Online.

Memorandum

To: Professor James Porter
From: Adam Banks, Andi Cumbo, Meg Fulton-Mueller, Naomi Igarashi, and Amy McAlpine
Date: October 30, 1999
RE: English 506 ? Computer Documentation Project Proposal

Objective & Purpose

To create a user-centered, basic procedural manual for developing a simple web page using Netscape Composer 4.6 (and subsequent versions), to be used by English Department faculty and graduate assistants ("faculty"). We believe that each faculty member should be able to create a web page in order to communicate effectively with students in CWRU's extensively wired campus. Our long-term goal is that faculty use this procedural document to design a web page that represents their unique interests and personality.

Pedagogically, the group's goals are to experience effective collaboration, learn to create a usable document, and develop strategies for teaching collaborative computer documentation projects. In order to use our knowledge next semester (and to overcome the problem of making these web pages usable), we recommend that the Spring English 398N class group project be developing documentation for loading web pages onto CWRU's I-drive.

Over the eight week project period, we will conduct a proficiency survey of faculty, a three-tiered usability test, and two follow-up surveys. At each stage of the project, we will consult with Professor Porter to assure our eventual success. During this time, Professor Porter will receive: a project proposal, a usability test report, and the final documentation and evaluation reports.

The following proposal details the project steps and research strategies.

Audience

Certainly, CWRU's faculty have varied computer competence levels. Through surveys distributed to the faculty, we hope to develop a clearer understanding of our audience's computer proficiency. We will assume that most of the English Department faculty are not familiar with Netscape Composer 4.5 and have not yet developed their own web page. However, we will assume that our audience is able to access email, word process and format a page. Using these criteria, we will develop documentation that leads readers easily through the web-design steps.

Document Content

The documentation will be published as a site on the Technical Writing web pages. At this time, we believe that the most effective way to use the documentation is to print it out prior to use. We will strive to keep the document short to facilitate printing. This will allow less experienced users to read the document completely without needing to have shift between two windows on the screen. More experienced users may shift from Netscape Composer to our document on-line.

Using a step-by-step approach, our readers will develop a web page with:

- links to email, syllabi, and the English Department
- CWRU headers (a requirement for posting a web

page on Aurora ? CWRU?s server)
- personal biography or course description (optional)

Our document will follow a linear format, offering a sample web page that the user may imitate or personalize. We will indicate that the web page may be loaded on the user's I-drive. Because I-drive is new and relatively unpredictable, we believe that the documentation for I-drive is best left for another project, possibly 398N's spring assignment.

Existing Documentation

Our document will provide links to existing documentation for faculty to explore more complex design options. Through an analysis of existing documentation, we will determine what these links will be. Also to inspire users to be creative with their web pages, the document will provide links to the English 398 syllabus and an existing web page, to be determined.

Usability Test Procedures

Our usability test has three stages. First, we will conduct a pilot test with members from our group, excluding the documentation developer who may be biased. During our pilot test, we will observe and take notes in order to improve the documentation.

Second, we will conduct a preliminary test with at least three participants. We hope to have English Department faculty possessing varied computer skill levels.

- Prior to the test, we will ask users to reveal their thoughts while working, to try to work through the

documentation independently, and that we will offer help only when asked. We'll assure them that we will offer them time for general feedback at the end of the test period.
- During these tests, we will have one person sit next to the document tester and take general notes and have another person overview the person's responses and take more detailed notes.
- After this test, we'll get additional feedback from users through both conversation and a brief exit survey.
- Using our results, we will further refine the documentation.

Third, a new group of at least three test participants will be observed using the same conditions as step two. From the three-tiered usability test, we will finalize our documentation.

Documentation Production Schedule (see Gantt chart)

The attached Gantt chart indicates the major stages of the project. The stages are:

1. write project proposal (week two), *
2. audience research survey (week two),
3. write documentation (week two, and on-going),
4. pilot test (week four),
5. preliminary usability test (week five),
6. final usability test (week six),
7. write usability test reports (week seven),*
8. finalize documentation and follow-up survey (week seven)*
9. write project evaluation (week eight).*

Items marked with an asterisk (*) will be submitted to Professor Porter. At each stage of the project planning, we hope to have Professor Porter's input and suggestions regarding our progress. During class time, we'll review our results to date and refine our strategies for the next week. Our usability tests will be conducted at the beginning of two class periods (November 1 and 8, 1999).

Our group has divided the tasks in order to maximize our efficiency:

- Fulton-Mueller - project proposal
- McAlpine - Gannt chart, usability test exit survey
- Cumbo - proficiency survey, existing documentation research
- Igarashi - documentation
- Banks - to be determined

Usability test result reports and final evaluations will be handled by the group. During the course of the project, we may further define the tasks according to need.

Potential Obstacles to Success

At this time, we think that the potential obstacles to our success include:

- failure to have usability test participants be objective thus skewing our findings,
- the inability to distribute document to faculty effectively,
- frustration caused while posting web pages on the I-drive or Internet, and
- faculty resistance to developing their own web pages.

Evaluation of Success

In the short run, if our usability tests reveal that our documentation enables users to create a basic web page in about 40 minutes without undue frustration, we will have accomplished our goal. At the end of the project, we plan to survey all the usability test participants again to see if they've used our documentation to create a web page and are satisfied with our product. From this survey, we'll have some standard for measuring user satisfaction. In the long run, we will assume that our project is successful if English faculty and graduate students use our document, or are inspired by our document, to create web pages in 2000.

Pedagogically, our success will be teaching the 398N collaborative computer project effectively.

Sample Application Letter - Research Position

Reprinted with the permission of Rebecca Mellis.

Rebecca Melton
1596 E. 115th St.
Cleveland, OH 44106

June 12, 2000

Dr. Stephen Haynesworth
Department of Biology
Case Western Reserve University
11900 Euclid Ave.
Cleveland, OH 44106

Dear Professor Haynesworth:

On March 10, 2000, I spoke with a representative from the SPUR Program sponsored by Case Western Reserve University Undergraduate Studies. We spoke at length about the requirements and characteristics of a successful student researcher. Because of my experience working in the Biology Laboratories coupled with my personal enjoyment of biology, I am confident that I have the ability to be a successful part of the SPUR student research team.

I have gained considerable lab experience through my present undergraduate assistantship. Currently, I work as an Undergraduate Biology Laboratory Assistant for the Genetics, Microbiology, Anatomy, and Physiology Labs under Mr. James Bader. Although my job responsibilities are diverse, I primarily focus on making media, inoculating cultures, preparing gram stains, operating lab equipment, and assisting other students with their laboratory assignments as they conduct their experiments. Additionally, I assist the instructor of the class by doing paperwork for the department and any other necessary and pertinent jobs.

Being afforded the opportunity to gain experience doing my own research project will give me a stronger basis of knowledge in my area of concentration. It will also provide first-hand knowledge of procedural methods, compiling data, and presenting finalized results of laboratory experiments. I am eager to learn from my fellow classmates and supervisors, and view this internship as a challenging and engaging activity. Both the research team and I will benefit from my curiosity and enjoyment of this area of study.

Enclosed are my resume, application materials, and statement of purpose. I look forward to meeting with you for an interview at your convenience. If you have any questions or comments, please do not hesitate to contact me. Thank you for taking the time to consider my application.

Sincerely,

Rebecca Melton
Enc.

Sample Resume - Construction Managment Internship

Reprinted with the permission of Jennifer K. Long.

Jennifer K. Long
2756 Hampshire Road, Apt. #2 • Cleveland Heights, Ohio 44106 • 216-333-3632 • jkl4@po.cwru.edu

Objective   To obtain a summer internship possibly leading to a full-time position that allows me to learn all aspects of financing and coordinating a construction project, from the initial concept to occupancy.

Education   Case Western Reserve University (CWRU), Cleveland, Ohio
Bachelor of Science in Civil Engineering (concentration in construction management)
Bachelor of Arts in English (concentration in film studies)
Date of Graduation: January 2001   GPA: 3.11/4.00

Gilmour Academy, Gates Mills, Ohio
Date of Graduation: May 1996   GPA: 4.13/4.00

Major Projects   Department of Civil Engineering, CWRU ('99-'00)
- Analyze the value engineering process for the Weatherhead School of Business project

College Scholars Program, CWRU ('99-present)
- Explore motivations of patrons of public art and healing power of public art

College Scholars Program, CWRU ('98-'99)
- Coordinate syllabus for study of city planning and extended learning trip to San Francisco

Skills   GeoPro, MS Front Page, MS Office 2000, Quickbooks 5.0, PCSTABL, Windows 98

Work Experience   Assistant, Teresa M. DeChant, Art Consultant, Cleveland, Ohio ('93-present)
- Help prepare and implement major public and private art installations and evaluations
- Prepare documentation including artwork listings, insurance information and project budgets

Intern, Osborn Engineering, Cleveland, Ohio (summer '97)
- Aid in updating structural drawings and assist with office documentation
- Utilize AutoCAD software

Intern, Historic Gateway Neighborhood Corporation, Cleveland, Ohio ('95-'96)
- Analyze the effects of urban gentrification on merchants
- Examine the impact of the public-private Gateway Sports Complex as an urban development plan designed to spur economic growth in downtown Cleveland

Volunteer Experience   Volunteer, AmeriCorps AQUACorps, CWRU (summer '98)
- Co-Create Kaleidoscope Jr., a biweekly environmental education program for ages 4-8
- Manage staff of seven for Kaleidoscope Jr. program

Volunteer and Staff, Youth Challenge, Cleveland, Ohio ('92-'98)
- Plan sports and recreational activities for children (ages 4-18) with physical disabilities
- Create the "YC Arts Fest," an annual art fair for seventy-five children and volunteers

Activities   Director of Membership, Student Turning Point Society, CWRU ('99-present)
Co-Chair, American Society of Civil Engineers Career Fair, CWRU ('99, '00)
Copy Editor, *The Observer*, CWRU ('97-'99)

Honors   Ohio Leadership, Peter Witt and Case Alumni Association Scholarships, CWRU
COOL Leaders Program, Campus Outreach Opportunity League
Sunshine Leadership Award, East Ohio Gas Company

Comments

Note that the resume is limited to one page. For a more detailed view see pages 102-104.

Sample Resume - Construction Internship (detail)

Reprinted with the permission of Jennifer K. Long.

Jennifer K. Long

2756 Hampshire Road, Apt. #2 • Cleveland Heights, Ohio 44106 • 216-333-3632 • jkl4@po.cwru.edu

Objective   To obtain a summer internship possibly leading to a full-time position that allows me to learn all aspects of financing and coordinating a construction project, from the initial concept to occupancy.

Education   Case Western Reserve University (CWRU), Cleveland, Ohio
Bachelor of Science in Civil Engineering (concentration in construction management)
Bachelor of Arts in English (concentration in film studies)
Date of Graduation: January 2001   GPA: 3.11/4.00

Gilmour Academy, Gates Mills, Ohio
Date of Graduation: May 1996   GPA: 4.13/4.00

Major Projects   Department of Civil Engineering, CWRU ('99-'00)
- Analyze the value engineering process for the Weatherhead School of Business project

College Scholars Program, CWRU ('99-present)
- Explore motivations of patrons of public art and healing power of public art

College Scholars Program, CWRU ('98-'99)
- Coordinate syllabus for study of city planning and extended learning trip to San Francisco

Skills	GeoPro, MS Front Page, MS Office 2000, Quickbooks 5.0, PCSTABL, Windows 98
Work Experience	Intern, Osborn Engineering, Cleveland, Ohio (summer '97) • Aid in updating structural drawings and assist with office documentation • Utilize AutoCAD software Intern, Historic Gateway Neighborhood Corporation, Cleveland, Ohio ('95-'96) • Analyze the effects of urban gentrification on merchants • Examine the impact of the public-private Gateway Sports Complex as an urban development plan designed to spur economic growth in downtown Cleveland
Volunteer Experience	Volunteer, AmeriCorps AQUACorps, CWRU (summer '98) • Co-Create Kaleidoscope Jr., a biweekly environmental education program for ages 4-8 • Manage staff of seven for Kaleidoscope Jr. program Volunteer and Staff, Youth Challenge, Cleveland, Ohio ('92-'98) • Plan sports and recreational activities for children (ages 4-18) with physical disabilities • Create the "YC Arts Fest," an annual art fair for seventy-five children and volunteers Director of Membership, Student Turning Point Society, CWRU ('99-present)

Activities	Co-Chair, American Society of Civil Engineers Career Fair, CWRU ('99, '00) Copy Editor, *The Observer*, CWRU ('97-'99)
Honors	Ohio Leadership, Peter Witt and Case Alumni Association Scholarships, CWRU COOL Leaders Program, Campus Outreach Opportunity League Sunshine Leadership Award, East Ohio Gas Company

Sample Reference Page

References -- Sara Vaughan

Mr. Luther N. Johnson
Manager
Information Services
Blues Online, Inc.
Indianapolis, IN 44834
(219) 476-4456
lgj@bluesonline.com

For the past two years, I have worked part-time for 20 hrs. per week as a web designer and information developer for Blues Online, Inc.

Prof. Luis Proenza
Vice President of Research
Dean of the Graduate School
Purdue University
West Lafayette, IN 47906
(765) 494-2604
lproenza@excel.purdue.edu

I served as a professional writing intern in Prof. Proenza's office (Summer 1996).

Prof. Johndan Johnson-Eilola
Department of English
Heavilon Hall
Purdue University
West Lafayette, IN 47907
(765) 494-1478
johndan@excel.purdue.edu

Prof. Johnson-Eilola was my instructor in two professional writing courses: Engl 424, Writing for the Computer Industry, and Engl 421, Technical Report Writing.

Prof. James E. Porter
Department of English
Heavilon Hall
Purdue University
West Lafayette, IN 47907
(765)494-3670
jporter@excel.purdue.edu

Prof. Porter was my instructor in one of my senior professional writing courses: Engl 515, Advanced Professional Writing

chapter 5

Principles

A useful way of organizing our work as professional writers is to think of our writing **practices.** We can focus on the genres of writing we produce: At times we write analytic reports, at other times we propose projects or actions, at other times we construct instructions for use or maintenance of a product, and so on. We can also focus on the ways that writing functions in an organization to allow (and sometimes inhibit) productive action: Writing may reflect a corporate culture's image, it may be used to attract investors, it may reflect/build consensus among groups from different segments of a company, it may intend to protect the company from lawsuits, and so on. We can further focus on how writers make good writing decisions.

The principles section of **Professional Writing Online** takes a reflective practices approach, one that focuses both on the production of effective professional writing and also on how and why a particular document meets the needs of a particular work situation. We use cases as a way to focus that tension between a particular writing action and taking good writing action. *PWO Projects* particularize our discussions of the principles of good professional writing at the same time as they keep alive the tension between action taking in specific situations and practicing good professional writing. The section of the Handbook that follows provides an overview of the table of contents for the Principles section of PWOnline. On the website itself, links to projects which use specific principles can be found under each section of the full table of contents.

Contents of "Principles"

The Principles section includes textbook-like discussions of the following topics:

Overview
- Rhetoric — Start With "Purpose" and "Audience"
- Understanding Purpose
- Understanding Audiences, Readers, Users
- The Complex Nature of "Writing"
- Evaluating Writing
- Ethics and Professional Communication
- Technology and Writing/Work in the 21st Century
- Collaboration and Team Writing
- Types of Documents (Genres)
- Terminology: What Is "Professional Writing"?

Understanding Readers
- Introduction to Readers
- What Readers Read
- How Readers Read
- How Readers are Often Classified
- When Readers Are Like Me (or not)

Social and Cultural Issues
- Ethics and Professional Communication
  - basic precepts of ethics
  - an introductory exercise in professional ethics — how documents can go wrong
  - ethical responsibilities for employees, employers, and
  - citizens (under development)
  - ethical guidelines for team work and project management
  - relations between ethics and law
  - copyright, plagiarism, and fair use of others' writing
  - representing others — race, ethnicity, gender,

               sexual orientation
          -   ethics editing checklist
- International Communication

Shaping Texts
- Introduction to Shaping Texts for Readers
- Markers for Readers
- Using ISIS as a Tool for Inspecting a Text
- Style, Audience, and the Arrangement of Information

Analyzing Workplace Writing
- Introduction to Analyzing Workplace Writing
  - Worksheet: Analyzing Workplace Writing Situations
  - Worksheet: Constructing a Document Plan
  - Worksheet: Responding to a Document Plan
- Representing Thinking Visually

Building Arguments
- Introduction to Building Arguments
- Types of Arguments
- Managing Information
- Arguing Visually
- Meeting Standards/Writing Specs
- Opening Statements
- Documents: Summarizing Information (abstracts, executive summaries)

Managing Projects
- Preparing Project Schedules
- Collaboration and Writing in Teams
  - Divide and Conquer
  - Shared Documents
  - Troubleshooting Team Problems
- Drafting and Revising Large-Scale Projects
- Preparing and Using Teamwork Logs
- Versioning Documents
- Organizing File and Directory Structures

Arranging Information
- Style, Audience, and the Arrangement of Information
- Thinking about Document Design
- Ordering Information for Understanding and Use
- Using Formats and Templates
- Constructing Visuals
- Designing with Color
- Using Citation Formats
- Building Tables

Research
- Introduction: Why Writers Need to Find and Use Information
- Methodology and Design
- Finding Sources that Already Exist
  - Searching Tools
    - Library
    - Web
  - Research Processes for Matching Sources and Info
    - Library
    - Web
  - Judging Sources
    - Library
    - Web
- Gathering and Analyzing New Information
  - Planning Informal Research
  - Informal Data Collection Techniques
    - Interviews
    - Questionnaires
    - Observations
  - Informal Data Analysis
  - Findings
  - Usability

Writing Reports
- A Generalized View of Reports

- Reporting in the Processes of Work
- Structured Writing
- Types of Reports
- Top-down Editing: A Structured Approach to Revising
- Developing Documents Across Technical Investigations

Writing Online
- Writing at the Computer
- Reading at the Computer
- Similarities and Differences between Print and Screen
- Designing Online Documents
- Communicating in Email, Chat, and MOO
- Designing Web Sites
- Designing Basic Multimedia

Usability Testing
- Introduction
- Overview of Methods for Document Testing
- Setting up a User Test (broad issues)
- Charting User Interactions with a Document
- Charting User Interactions with a Product
- Aids from a Guide for Informal Usabilty Testing
- Worksheet: Question Posing and Selecting
- Example Observation Sheets
- Worksheet: Analyzing Data

Style
- Style Overview
- Types of Revising and Editing
- Style, Audience, and the Arrangement of Information
- Citing Sources and Figures in the Report Text
- Style Exercise
- Style and Editing Activity
- Talking Heads

The Principled Perspective

At times the discussions in Principles can be quite detailed, as you expect from a textbook on Business or Technical Communication. This is true when we are discussing writing practices that relate to good habits for writers -- for example, the habit of analyzing writing situations. Because workplace writing is a situated activity, good writers habitually analyze how the roles that writing plays, both historically and now, fit into an organization's processes of work, audiences, and corporate identities. PWOnline's Principles section examines the components of this good writing habit deliberatively and offers a worksheet to use in building writerly habits of analysis. (Note: Worksheets are also tailored for projects when appropriate.)

Principles is also quite detailed in its discussion of professional writing as rhetoric, understanding readers, ethics, shaping texts for readers, building arguments, managing projects, understanding structures for information, style, and conducting research. It also offers the traditional principles behind reports as structured writing and particularizes that approach with discussions of traditional workplace genres (letters, memos, policies, resumes and so on) with practical advice about structured editing. But, Principles does not stop with structured reporting, adding detailed discussions of online writing and usability.

From a principles perspective we find PWOnline particularly rich and flexible. If a writing situation or course calls for more emphasis on structuring information than on understanding audience, such an emphasis is possible. It is also possible to emphasize (or de-emphasize) argument, style, arrangement, genre, writing process, technology, and even rhetoric itself. Such is the fate of Principles -- to wait patiently for recognition even though they are our bedrock.

chapter 6

Resources

The Resources section of the PWOnline website provides links to a variety of web resources that professional writers and professional writing instructors use as they conduct the online writing tasks promoted by Professional Writing Online. It provides web links that assist in hunting for jobs, doing research for business or technical reports, developing online mateials (e.g., finding graphics for a web site), and communicating in online work cultures.

Writer's Web Links

This section also serves as a quick reference for sources that are explained elsewhere in PWOnline and it serves, too, as a reminder that the web can be of great assistance to professionals as they write. Each reference includes a review of its uses. So, for example, if you are searching for free click art for your web site, you can find helpful sources among the graphics and visuals resources. If you are looking for assistance in constructing tutorials, the online documentation and the usability resources are invaluable.

The Resources section includes pages with links to the following types of information:

- instructor resources
- search engines
- searching for information
- corporate web sites
- web design resources
- graphics and visuals
- online documentation
- internet use statistics
- usability resources
- design tips and templates
- professional writing
- citing sources
- job and career resources
- professional writing programs

Writer's Exercises and Worksheets

Professional Writing Online also includes a number of activities and exercises designed to help guide you through specific writing problems. The following table lists the exercises in PWOnline.

Table 6-1: Writer's Exercises in PWOnline

Exercise	Follow the link from	In Topic	Section
Evaluating the Design of Online Documentation	Exercise available directly from the "Projects" sidebar in Projects	Evaluating the Design of Online Documentation	Projects
Preliminary Analyses of Complaints	Exercise: Preliminary Analyses for Complaint Letters	NCAA Bat Standards	Projects
Exercise: Writing Questions that Sort and Target	Gathering & Analyzing New Information	Research	Principles
Exercise: Interpreting Data	Gathering & Analyzing New Information	Research	Principles
Exercise: Open and Closed Questions	Gathering & Analyzing New Information	Research	Principles
Exercise: Planning Informal Field Research	Gathering & Analyzing New Information	Research	Principles
Exercise: Questionnaire Revision	Gathering & Analyzing New Information	Research	Principles
Exercise: Writing Good Questions	Gathering & Analyzing New Information	Research	Principles
An Introductory Exercise in Professional Ethics	Introductory Exercise in Professional Ethics	Social and Cultural Issues	Principles
Correct the Following Stylistic Problems	Style and Editing Activity	Style	Principles
Editing for Style Exercise	Style Exercise	Style	Principles

Professional Writing Online also includes a number of worksheets to help you plan and organize projects. The following table lists the exercise in PWO and identifies their location on the website.

Table 6-2: Writer's Worksheets in PWOnline

Worksheet	Follow the Link from	In Topic	Section
Airbag Safety Letter at Vanguard	Airbag Letter Planning Worksheet	Airbag Case	Projects
Letter Purpose Analysis Chart	Letter Purpose Analysis Chart	Airbag Case	Projects
Customer Complaints at Allied Mutual	Allied Letter Planning Worksheet	Allied Mutual Insurance Case	Projects
Letter Purpose Analysis Chart	Letter Purpose Analysis Chart	Allied Mutual Insurance Case	Projects
Job Banks Preliminary Analysis	Project Analysis Worksheet	Job Banks	Projects
Analysis for National Electric Visuals Project	Project Analysis Worksheet	National Electric	Projects
NCAA Bat Standards – Project Planning	Planning for this Project	NCAA Bat Standards	Projects
Building a Usability Testing Logsheet	Usability Testing Logsheet	Software Learning Initiative	Projects
ISIS for Documentation Analysis	ISIS Worksheet for Analyzing Documentation	Software Learning Initiative	Projects
Big 1 Writing Situation	Worksheet for Case Analysis	The Big 1 Case	Projects
United Drill Writing Situation	United Drill Project Worksheet	United Drill	Projects
Analyzing Writing Situations	Analyzing Workplace Writing Situations	Analyzing Workplace Writing	Principles
Responding to Planning Documents	Responding to a Document Plan	Analyzing Workplace Writing	Principles
Worksheet for Starting Doc Projects	Designing and Writing Instructions – follow link to "worksheet" in introductory paragraph	Instructional documents	Principles
Worksheet: Analyzing Data	Worksheet: Analyzing Data	Usability Testing	Principles
Worksheet: Question Posing and Testing	Worksheet: Question Posing and Testing	Usability Testing	Principles

Instructor's Resources

Whether you are teaching studetns in a college classroom, running a training seminar at work, or teaching yourself at home, PWOnline provides the on- and off-site resources that will be useful to your instructional efforts. You'll find links to all of our instructor resources in the Resources chapter. Just follow the link to "Instructor Resources."

On-site resources include links to:

- how to use professional writing online
- instructor notes on projects
- chat rooms
- PWOnline e-mail discusion list

Off-site resources include links to:

- e-mail discussion lists
- professional writing programs – undergraduate, masters, doctoral
- course syllabuses

Interactive Features and Functions

PWOnline offers facilities for interactive learning, including:

- chat spaces where you can talk to people in your class or people in other classes around the world
- newsgroups for general information about PWOnline (as well as newsgroups for your specific class, if your teacher chooses to set one up)
- online forms distributed throughout the site that allow you to begin work online and email work to your teacher or students in your class for comments and feedback

Working in Newsgroups

Newsgroups in PWOnline provide a central "space" for users to send and receive messages about different topics. Newsgroups are usually "asynchronous," meaning that users usually post and read messages at different times (unlike chat spaces, where users discuss topics at the same time, "live"). As you might guess, newsgroups allow people with different schedules to engage in discussion—they're sort of like email (which you might use in PWOnline exercises and worksheets, as well as numerous other areas), except that they're more public and can be thought of as having a centralized location.

PWOnline supports three different types of newsgroups (paralleling the PWOnline chat spaces):

- Class-specific newsgroups: assigned to a single class (these must be requested by instructors). These spaces will usually only have people from your specific class in them.

- Issue-specific newsgroups: set up to cover different projects and topics in PWOnline as needed. These can be used by any PWOnline user, and will probably include people outside of the class you're taking.

- Coffeehouse newsgroups: open to any topic discussion for any PWOnline user. People here could be talking about anything, not necessarily limited to PWOnline. But: Avoid harassment of any type (sexual, racial, classist, etc. It's fine to engage in heated discussion, but be respectful of each other. The management reserves the right to remove users and, in extreme cases, revoke PWOnline accounts.

There are currently three newsgroups up for beta testing.

- Newsgroup 1
- Newsgroup 2
- Newsgroup 3

PWOnline authors and Allyn & Bacon are not responsible for the specific actions and content of users in these spaces.

Working in Chat Spaces

Chat spaces provide you with an online place to discuss issues, problems, and questions about professional writing and PWOnline assignments with other users. In some cases, these users will be students in your own class, who may be using the chat space to meet outside of class at a scheduled time (with different students in different locations, on and off campus). In other cases, you'll join a chat space to talk to students in different classes around the country or even around the globe. In some cases, you may even use a chat space to talk to people in your class even though you're all sitting in the same computer lab working. Unlike newsgroups, chat spaces happen in realtime, with users exchanging messages with each other as fast as they can type.

You might wonder why anyone would use a chat space rather than send an email message or simply speak out loud to another person. The answer is simple: different media offer different ways of interacting with people. In newsgroups, users typically have a lot of time to think about their messages, reading and re-reading them if necessary, because discussion happens over long periods of time (hours or even days), with each user posting messages that may be read much later. In a face-to-face classroom conversation, in most cases only one person can speak at one and, in most cases, the instructor talks more than anyone else. In a chat space, though, the numerous people can communicate at the same time with less chance of blocking out each others' contributions.

Working in chat spaces can take some getting used to—the pace is often very fast, and messages written by other users can quickly pile up on your screen. Typing speed can also be an issue—people who can type quickly have a "louder voice" than people who type slowly.

Note: Users can create temporary rooms for collaborative work by clicking the "create room" button in the lower-right corner of the i-chat window. Use temporary rooms for team meetings, brief conferences, etc.

PWOnline will eventually provide three types of chat spaces (paralleling the PWOnline newsgroup structure):

- Class-specific spaces: assigned to a single class (these must be requested by instructors). These spaces will usually only have people from your specific class in them.

- Issue-specific spaces: set up to cover different projects and topics in PWOnline as needed. These can be used by any PWOnline user, and will probably include people outside of the class you're taking. Currently, only the Corporate Web Project Chat Space is online (to be used, obviously, with the PWOnline Corporate Web Project).

- Coffeehouses: open to any topic discussion for any PWOnline user. People here could be talking about anything, not necessarily limited to PWOnline. But: Avoid harassment of any type (sexual, racial, classist, etc. It's fine to engage in heated discussion, but be respectful of each other. The management reserves the right to remove users and, in extreme cases, revoke PWOnline accounts.

For now, we're limited to a single, Coffeehouse space for the prototype: try out the prototype coffeehouse.

PWOnline authors and Allyn & Bacon are not responsible for the specific actions and content of users in these spaces.

Use the form to send us your questions and comments, or e-mail us at pwonline@addison.english.purdue.edu.

Feedback

We would like PWOnline to be an interactive and lively site that is responsive to your needs as users. We want to keep the site updated and current and to provide information that is useful to your teaching and learning. To that end, we would welcome three types of feedback in particular:

- **Editorial Fixes** — please notify us if you find broken or obsolete links or other errors anywhere on the site.

- **Advice/Reactions** — please tell us how you like various sections, what works and what doesn't.
- **Contributions** — please send us material and URLs you think would be useful on the site. Eventually, we will install a means for you to submit syllabi, projects and cases, sample documents, and other materials for possible inclusion in PWOnline.

We are particularly interested in developing a resume and application letter gallery.

about

the Authors

Jim Porter (PhD, University of Detroit, 1982) has taught technical and business writing at the university level since 1982. He is currently Professor of English at Case Western Reserve University, where he serves as Director of Technical & Professional Communication. Between 1988 and 1999, Jim worked at Purdue University, where he taught courses in the undergraduate professional writing program and in the Rhetoric/Composition PhD program. His books include *Audience and Rhetoric* (Prentice Hall, 1992), *Opening Spaces: Writing Technologies and Critical Research Practices* (with Pat Sullivan, Ablex, 1997), and *Rhetorical Ethics and Internetworked Writing* (Ablex, 1998). His co-authored book with Pat Sullivan won the 1998 NCTE Award for Best Book in Technical and Scientific Communication; *Rhetorical Ethics* won the Computers & Writing Best Book award in 1999; and twice he and Pat Sullivan have won the NCTE award for Best Article in Theory/Philosophy in Technical and Scientific Communication (1994, 1997). From 1991-1993, Jim served as a freelance consultant to the Usability Group at Microsoft Corporation.

Patricia Sullivan (PhD, Carnegie Mellon University, 1985) is Professor of English and Director of the Rhetoric/Composition PhD program at Purdue University. She teaches undergraduate courses in professional writing and graduate courses in methodology, rhetoric theory, visual design, and professional writing. She has authored *Opening Spaces* (Ablex 1997) with Jim Porter and co-edited *Electronic Literacies in the Workplace* (NCTE 1996) with Jennie Dautermann. Pat is a four-time winner of an NCTE award for one of the year's best publications in Technical and Scientific Communication.

Johndan Johnson-Eilola (PhD, Michigan Tech University, 1993) works as the Director of the Center for Excellence in Communication and an Associate Professor of Technical Communications at Clarkson University.

He has previously served as the director of technical and business writing programs at Purdue University and New Mexico Tech. His published work has been awarded the Computers and Writing Best Article Award (1996) and the Technical Communication Quarterly Best Article Award (1997). He is also the author of *Nostalgic Angels: Rearticulating Hypertext Writing* (Ablex/JAI Press) and the forthcoming *Designing and Writing Hypertext* (Houghton Mifflin).

Other Contributors

People who contributed materials to the site:

- Teena Carnegie
- Barbara Couture
- Bill Hart-Davidson
- Scott Jones
- Amy C. Kimme Hea
- Tim Krause
- Colleen Reilly
- Jone Rymer
- Mark Schaub
- Michele Simmons
- Melinda Turnley

People who worked on construction of the site:

- Stuart Blythe
- Laurie Cubbison
- Kevin DePew
- Bill Hart-Davidson
- Amy C. Kimme Hea
- Erin Karper
- Amy McAlpine
- Colleen Reilly
- Julia Rosenberg
- Carlos Salinas
- Michele Simmons
- Julie Woodford

Index

A
Airbag Safety Case 36–37
Analysis of Writing Practices 26–28
Analyzing Professional Contexts 28–31

B
Big 1 Case 45–47

C
Chat Spaces 118–120
Copyshop Copying Policy Case 38–39
Corporate Web Project 42–45
Customer Complaints at Allied Mutual 32–35

D
Designing Online Documentation 21–24
Discussion questions
  Airbag Safety 39
  Analyzing Professional Contex 33
  Big 1 Case 46
  Copyshop Copy Policy Case 40
  Corporate Web Project 45
  Customer Complaints at Allied Mutual 35
  Designing Online Information 23
  E-Commerce Project 70
  Employment Project 50
  Evaluating Online Design 24
  Job Banks Project 66
  National Electric 68
  Software Bug Case 65
  Software Learning Initiative 21
  Technology Access Memo/Report 57
  The Kmart Case 48
  United Drill Case 43

E
E-Commerce Project 70
Employment Project 50–51
Evaluating the Design of Online Documentation 24–26

F
Feedback 119

I
Instructors' Resources 116
Interactive Features 116

J
Job Banks Project 65–68

K
Kmart Case 48–50

L
Lester Crane Case 51–55

N

National Electric: Creating Visuals 68
NCAA Baseball Bat Standards: 59
Newsgroups 117–118

O

Online Class Schedule 9

P

Principled Perspective 112
Project-oriented Approach 5–6
PWOnline Web Site
    Web Site URL 1
    Writer's Web Links 113–117

R

Reflective Practices Approach 107

S

Samples
    Letters
        Application Letter - Research Position 99–100
        Request for Deposit Refund 74–75
        Revised Request for Deposit Refund 76–77
    Memos
        Announcement of Policy Change 78
        Project Planning Memo 81
        Revised Announcement of Policy Change 79
    Reference Page 105
    Reports
        Corporate Web Project Proposal 86–88
        Corporate Web Recommendation Report 89, 89–92
        Documentation Project Plan 93–98
    Resume
        Construction Position 101, 102-104
Sitemap 15

Software Bug Case 62
Software Learning Initiative 19

T

Tables
    Projects in Professional Writing Online 17
    Writer's Exercises in PWOnline 115
    Writer's Worksheets in PWOnline 114
Teaching Scenarios
    Using PWOnline to Support International 11
    Developing Distance Collaboration 10
    Tailoring Projects for Local Situations 10
    Teaching from a Documentor Principles 7–8
    Teaching from a Project-Oriented Approach 6
    Using PWOnline in the Distance Education 8
    Using PWOnline in Traditional Classrooms 9
Technology Access Memo/Report 56–59

U

United Drill Case 40–41
Users' Handbook
    Using the Handbook 2